QUAKER QUICKS

Quakers and Science

What people are saying about

Quakers and Science

This book makes a strikingly original contribution to the science-and-religion debate. Through a series of bite-sized biographies Helen Holt explores the distinctive approaches that Quaker scientists have brought to their scientific work. Emphasising shared commitments to social justice, pacifism, experience and the Inner Light, Holt paints compelling and human portraits of both Quakerism and science. This book stands out as an important milestone in studies of science and religious faith.
Mark Harris, Professor of Natural Science and Theology, University of Edinburgh

My formal scientific education ended with O-levels in 1955 but I have been aware for most of the 40-plus years that I have been a Friend that individual Quakers have over the years made significant contributions to the advancement of scientific knowledge and increasingly aware of the fact that the Quaker approach to religion is very much in tune with the scientific method. How good it is, then, to have this spelled out for me in this little volume, which I found to me very readable, informative and, beyond that, inspiring. I hope that it will be widely read and not only by Quakers. It should have a prominent place in Quaker outreach work and, of course, in every meeting house library.
Phil Lucas, ex-clerk of Quaker Life Central Committee

A highly insightful account of 10 prominent Quaker scientists who have exhibited a wide range of diverse and creative reactions to the issues of science and religion, as well as how to be a socially responsible human being. Together they present

Quakerism as an attractive and open-minded force for good in the world.
Eric Priest, FRS, St Andrews University

This clear and helpful book explains how Quakers have brought together science and religion, not without some struggles but often joyfully and productively. Giving short biographies of a range of Quaker scientists from the early twentieth century up to today, Helen Holt shows how scientific work shaped their religious understandings and how their religious commitments, especially to pacifism, affected their careers.
Rhiannon Grant, author of *Hearing the Light* and other Quaker Quicks books

Quaker Quicks

Quakers and Science

Helen Holt

CHRISTIAN ALTERNATIVE BOOKS

Winchester, UK
Washington, USA

JOHN HUNT PUBLISHING

First published by Christian Alternative Books, 2023
Christian Alternative Books is an imprint of John Hunt Publishing Ltd.,
No. 3 East St., Alresford, Hampshire SO24 9EE, UK
office@jhpbooks.com
www.johnhuntpublishing.com
www.christian-alternative.com

For distributor details and how to order please visit the 'Ordering' section on our website.

ISBN: 978 1 80341 139 2
78 1 80341 140 8 (ebook)
Library of Congress Control Number: 2022933593

A CIP catalogue record for this book is available from the British Library.

Design: Stuart Davies

UK: Printed and bound by CPI Group (UK) Ltd, Croydon, CR0 4YY
US: Printed and bound by Thomson-Shore, 7300 West Joy Road, Dexter, MI 48130

Contents

Author of *Mysticism and the Inner Light in the Thought of Rufus Jones, Quaker* (Brill, 2022). ISSN 1566-208X

For my sons, Calum and Andrew

Preface

Whether or not we are religious, and whether or not we are scientists, religion and science have shaped our everyday lives in numerous ways. In spite of the fact that these two great endeavours are often portrayed as being in conflict, they are in practice often tightly interwoven. We might be given the latest gadget for Christmas, for example, or argue for the need to donate COVID vaccines to developing countries on the basis of religiously inspired ideas of justice and compassion. The academic study of the relationship between science and religion began in earnest in the 1960s and is a fascinating field. It investigates topics as varied as what spirituality might look like in a digital age, the problem of how God might interact with the physical world, and the implications for free will arising from the discovery that our decision-making is affected by the bacteria in our gut.

Quakers have much to contribute to discussions about this relationship. This is partly because they have such a long and strong tradition of encouraging scientific pursuits. When George Fox and others founded Quakerism (more formally known as the Religious Society of Friends) in the 1650s, scientific discoveries were transforming the way the world was understood: in the three decades either side, William Harvey revealed the role of the heart and arteries in circulation, Robert Boyle divorced chemistry from alchemy, and Isaac Newton discovered the laws of motion. Clearly, scientific knowledge has deepened and widened enormously since then, and scientists (a term first used in 1833) have shifted from having an amateur to a professional status. But if we think of science fairly loosely as an investigation into the nature of the physical world involving observation and experiment, then Quakers have always been involved: George Fox was clear about the benefits of studying nature in the seventeenth century, Quaker amateur botanists were widely respected in the eighteenth, Quaker industrialists made scientific discoveries in the nineteenth, and eminent Quaker scientists made key contributions in fields ranging from physics to psychology in the twentieth.

The Quaker contribution to discussions about science and religion is also down to the fact that Quakers have a distinctive approach to religion. Quakerism has its origins in Christianity, but it has traditionally placed more emphasis on deeds than creeds, and has always insisted that there is 'that of God' in every person, with this point of contact referred to as the Inward or Inner Light. One consequence of this is that the questions Quakers have grappled with regarding the relationship between science and religion are often not primarily theological (for example, why a loving God would allow the suffering involved in the 'survival of the fittest') but ethical, and, furthermore, these questions demand a response. Victorian Quakers argued that vivisection should be banned because it was a vice that hindered people from hearing the guiding voice of God, for example, and, as pacifists, a number of twentieth-century Quaker scientists went to extraordinary lengths to make sure that their research would not be used for the purposes of war.

The aim of this short book is to explore how Quakers have seen the relationship between science and their faith, and to look at how some individual Quaker scientists wove together their passions for both in their lives. For reasons of space, it does not discuss Quaker medics, nor the Quaker pioneers who in the early nineteenth century made such compassionate and ground-breaking changes to the care of those in asylums. It is split into three chapters. The first chapter gives a broad overview of how Quakers viewed and practised science from the movement's beginnings until the Manchester Conference of 1895, an event that saw British Quakerism (and a significant strand of American Quakerism) begin to move towards a deliberate engagement with science and other aspects of 'modern thought'. The second chapter adopts a different approach and introduces ten significant Quakers—eight practising scientists and two influential Quaker reformers who studied and taught science. The final chapter makes a few comments about the resonances between Quakerism and science.

In spite of the richness of the relationship between Quakerism and science, surprisingly little has been written about it. Much of what has is in academic journals, which are not readily available to the general public. Furthermore, many of the books written by the scientists themselves are no longer in print, although some are available through Internet Archive (a free resource available at www.archive.org). For those who want to follow up on some of the topics, I've listed a couple of the more accessible books under 'further reading' at the end of each section, and there is a fuller bibliography at the end of the book. More generally, I hope that readers without much knowledge of Quakerism will be able to pick up some of the basics of Quaker beliefs and practices along the way, but there is a short Glossary at the end, and www.quaker.org.uk is a useful website for background information.

Further reading

A very readable introduction to the relationship between science and religion is Alister E. McGrath, *Science and Religion: A New Introduction* (Wiley Blackwell, 2010).

Good short introductory books on Quakerism include Pink Dandelion, *The Quakers: A Very Short Introduction* (OUP, 2008), and a number of books in the Quaker Quicks series, for example, Rhiannon Grant, *Quakers Do What! Why?* (Christian Alternative, 2020). A more comprehensive volume, covering a range of topics, is Stephen W. Angell & Pink Dandelion (eds.), *The Oxford Handbook of Quaker Studies* (OUP, 2015).

Acknowledgements

I am very grateful to Phil Coutts, Rhiannon Grant, Mark Harris, Calum Holt, Phil Lucas and Eric Priest for reading earlier versions of this book and providing useful comments from a range of perspectives. Any errors remain, of course, my own.

I'd also like to thank Sue Proudlove and Tony Wilkes, managers of Edinburgh Quaker Meeting House, for digging out books for me in the midst of library renovations and COVID restrictions.

Chapter 1

Quakers and science before 1900

A number of writers from the 1920s onwards enthusiastically trumpeted the success of Quaker scientists, claiming that Quakers had secured over forty times their due proportion of Fellows of the Royal Society. Subsequent more careful analysis by Geoffrey Cantor has shown this claim to be both exaggerated (non-Quaker scientists with common Quaker surnames were often counted as Quakers), and in a sense misleading as a reflection of scientific prowess (until the mid-nineteenth century, the Royal Society was something of a gentleman's club rather than an indicator of scientific distinction). Be that as it may, there is no doubt that Quakerism has produced many keen amateur and eminent professional scientists, so we'll look in this chapter at some of the historical reasons for this and for the particular scientific interests that Quakers had prior to the twentieth century.

In many ways, Quakers were historically almost corralled into practising science by various restrictions. As non-conformists they were excluded from many professions, as pacifists they objected to military service, and their refusal to take oaths (on the basis that it implied a double standard of truth-telling) excluded them from holding office. This meant that Quakers did not partake in professional life, the military or Parliament for the whole of the eighteenth and much of the nineteenth centuries. Not only that, but Oxford University required students to subscribe to 39 articles of faith prior to the mid-nineteenth century, while Cambridge allowed non-Anglicans to enter but not to graduate, meaning that university-level education in England was limited for Quakers.

The flip side of these restrictions was that many of the

occupations that were open to and embraced by Quakers involved science. The mine-owner Robert Were Fox (1789–1877), for example, published 60 papers on topics related to his industrial and commercial interests, including on underground electric currents and the use of steam engines in mining. In industry, the Darby family established iron works in Shropshire in the eighteenth century, and the Cadburys, Frys and Rowntrees were famous for their chocolate manufacturing in the nineteenth. In order to be successful in such businesses, these Quakers had to keep up to date with relevant scientific and technological developments. Away from industry, a number of Quakers developed a reputation as traders of plants and seeds. Peter Collinson (1694–1768), for example, became the foremost importer of botanical specimens from North America in the mid-1700s. He introduced nearly 200 plant species into Britain, and also shipped books and animals, and, apparently, even the hand and arm of a 'maremaid' from Brazil. Other wealthy Quakers became internationally recognized as collectors of shells, minerals and plants. Thomas Hanbury's garden at his La Mortola estate in Menton, Italy, for instance, was one of the most celebrated gardens in the world and was visited by many serious botanists.

A further restriction that indirectly led to an interest in science was that Quakers were limited in their choice of leisure activities, although here the restrictions came not from wider society but from Quakers themselves, who prohibited music, card games and novels until the late nineteenth century. Botany was one of the few pastimes that was encouraged, and the Quaker feminist, philanthropist and science writer Priscilla Wakefield (1751–1832) was not alone in insisting that botany was not a frivolous amusement but an 'innocent enjoyment' that required patience and perseverance and helped train the mind and eye. This attitude was strikingly apparent at Bootham School in York, where the head John Ford (1801–1875) founded the Natural History Society in 1834, in part from a desire to occupy his pupils after school: one report to the Friends' Education Society enthused that participation in natural history channelled the energy of students away from 'idle, frivolous, or pernicious pursuits ...[and] from the society of the gay and trifling'![1] The Society amassed specimens, awarded prizes, published its own journal, and even had its own telescope. For many pupils the Society formed a conduit to a scientific career, and indeed the school became known as 'that great nursery of Quaker botanists'.

Scientific pursuits were not just an occupational necessity or acceptable pastime, however. Many Quakers felt that the study of nature was a spiritual exercise that brought them closer to God. The seeds for this conviction were perhaps sown by George Fox and other early Quakers, who believed that Adam had possessed a true knowledge of creation that was lost in the Fall but could be regained in a state of grace. Fox writes of a religious experience in which he came into the paradise of God and found himself in a state of innocence. In this state, he continued, 'all the creation gave unto me another smell than before, beyond what words can utter... The creation was opened to me; and it was showed me how all things had their names given them according to their nature and virtue'.[2]

While Fox moved from a religious experience to a transformed knowledge of creation, later Quakers found that studying creation led to spiritual growth, solace or wonder. Thus the Bootham-educated naturalist George Stewardson Brady (1832–1921) advocated the study of natural history because, in contrast to the tumult and din of 'Salvation Armies' and the spiritual intoxication of revivalist meetings, nature worked quietly on the soul, initiating changes that were both deeper and more long-lasting. For James Backhouse (1794–1869), who owned an extensive nursery in York and travelled to Australia offering religious instruction to convicts on behalf of the Quakers, botany provided a source of beauty and spiritual refreshment on his expeditions that contrasted with the 'dark shadow' that fell on the moral world around him. And when a contributor to the Quaker journal the *Friend*, identifiable only by his initials H.R., wrote a series of articles on astronomy in 1843–1844, he concluded that contemplating the stars could not fail to 'raise our views and conceptions of the greatness of the Deity'.[3]

Interestingly though, some Quakers perceived that in tension with this view of the study of science as almost a spiritual exercise was the possibility that one could have too much of a good thing, such that the practice of science might be detrimental to their spiritual life. John Rutty (1698–1775) studied, among other things, the influence of weather conditions on disease. While he saw his research as enabled by God and acknowledged that science could lead to a state of grace if practised with humility, he also prayed to be forgiven for idolising nature and spending too much time studying it. And John Barlow (1815–1856), who trained in veterinary medicine at Edinburgh, reflected during his final months that, although he had dearly loved science, he had followed it with too exclusive a devotion and with a care for his scientific reputation.

As a final point, it is worth mentioning that, although Quaker women did not receive equity with men in science education, they had more exposure to science than women did generally. The Quaker Dr Thomas Pole, for example, gave a series of scientific lectures in 1802 and stated that they would be adapted to persons of both sexes, as he deemed women to have been excluded from opportunities for scientific improvement. Furthermore, women were exposed to science because they played a prominent role in the Society generally. Because Quakers believe that there should be no separation between the religious, social and political realms, women were involved in a wide variety of issues, including assisting with the family business, which, as we have seen, sometimes meant they needed to be informed about scientific and technological advances. One notable example is the astronomer Maria Mitchell (1818–1889). Mitchell grew up in the whaling town of Nantucket, Massachusetts, helping her father in the makeshift observatory on the roof of their tiny house. One night in 1847 she discovered a new comet, which was later named 'Miss Mitchell's comet' in her honour. She subsequently accepted a position at Vassar College, New York, and in so doing became America's first woman professional astronomer.

This short survey has touched on some of the reasons Quakers embraced science in the first 250 years of the Society's existence. But how did the practice of science and scientific knowledge relate to the fundamental Quaker belief that there was 'that of God' in everyone, meaning that all people should be treated equally and with respect?

One way in which this idea found expression was a concern with the moral questions raised by scientific discoveries, especially in relation to ethnology and anthropology. The British doctor James Cowles Prichard was raised as a Quaker, although he later converted to Anglicanism, and has been hailed as one of the fathers of anthropology. In his *Researches into the Physical History of Man* (1813) he argued strenuously for monogenism, the unity of the human races. The preface to the book notes that he wanted to reconcile the observed diversity of races with the biblical record that teaches that all humans descended from common parents. He thus used linguistic, cultural and physical evidence to establish affinities between different human groups. Crucially, monogenism has the natural consequence that all aborigines should be treated as brothers and sisters. Thomas Hodgkin, an Edinburgh-trained doctor, thought similarly, and was active in founding the Aborigines' Protection Society in 1837, which aimed to collect information from travellers about the conditions of 'defenseless or uncivilised tribes' and to persuade Europeans to behave humanely towards them.

Statistics, too, were put to use in the Quaker war against injustice. Regarding the slave trade, for example, anecdotal accounts of suffering were important, but prominence was given to statistical evidence because it added scientific credibility and transcended the experience of individual travellers. Thus an article in the Quaker journal the *Friend* in 1843 gave a historical account of the slave trade and cited the number of slaves transported from Africa for each trading nation, showing that in 1788 Britain was responsible for half of the total number.

A further ethical concern was for the wellbeing of animals. A precedent had been set as early as 1772, when John Woolman, the American one-time tailor and abolitionist, insisted on making long journeys on foot when he visited Britain rather than ride in a stagecoach. Stagecoaches, he wrote with obvious incredulity and disapproval, frequently go upwards of one hundred miles in twenty-four hours. He was aggrieved not only that post-boys sometimes froze to death on winter nights, but also that horses were being driven to death. 'So great is the hurry in the spirit of this world', he lamented, 'that in aiming to do business quickly and to gain wealth the creation at this day doth loudly groan'.[4]

In the late nineteenth century, this concern was extended to vivisection. Although Quakers used the vocabulary and arguments of other Victorian Christians, they brought a unique perspective, namely the peace testimony (on which more in Chapter 2), characterized by pacifism and non-violence. Quakers began to debate who or what should be protected from violence, and mounted arguments against vivisection on a number of fronts. They argued, for example, that vivisection should be prohibited because it was like slavery, in that both involved the mistreatment of a weaker group. They also drew on the Bible, quoting verses related to the peaceable kingdom (Isa. 11:1–9) and dominion (Gen. 1:28), which they interpreted as divine commands to care for animals. Interestingly, they also argued that vivisection damaged virtue. On an individual level, the men involved in vivisection could not be trusted to make a decision about whether it should be continued, the implication being that vivisection hampered a person's ability to be guided by God. Not only that, but society would be damaged, as entire groups would buy into this system of violence and be infected by vice.

Not surprisingly, Quaker participation in science and the concerns generated by it were reflected in Quaker education, both at home and in Quaker schools. Two prominent Quaker writers who produced popular scientific educational texts for children and young adults were Priscilla Wakefield, mentioned above, and Maria Hack (1777–1844). At the time these women were writing, the work of William Paley was prominent as a way to reconcile new discoveries in science with religion. Paley was an English clergyman, and his *Natural Theology or Evidence of the Existence and Attributes of the Deity* (1802) became famous for its 'Watchmaker analogy' (as lampooned in Richard Dawkins' 1986 book *The Blind Watchmaker*). If we came upon a watch on a heath, Paley argues, we would deduce that the watch must have had a maker; likewise, the complex design apparent in nature, for example, in the human eye, implies that there must be a divine designer. Both Wakefield and Hack built on the work of Paley. Their books, like many others of that era, took the form of a conversation between a child and mentor. In Hack's *Harry Beaufoy; Or, The Pupil of Nature*, the fictional Harry is in the garden with his parents one evening, spellbound by the moon, which was suspended 'like a lamp of gold over the dark tufted trees'. While the heavens do indeed declare the glory of God, acknowledges his father, it is also important for Harry to observe God's creatures and find evidence for himself of design in their instincts and needs: 'If once we prove design, you know Harry, we are sure that there must be a contriver', he says.[5]

In Wakefield's books, children are taken to manufacturing sites, and the science behind industrial processes is explained to them. But the importance of ethical behaviour and Quaker practice is never far from the surface. In *Mental Improvement*, there is an educational family conversation about sugar cane, in which Henry, Cecelia, Sophia and Mr and Mrs Harcourt discuss its genus (*Triandria digynia* class) and characteristics (a woolly down, longer than the flower that encloses it). Mr Harcourt goes on to explain that because of the lack of horses in the West Indies, almost all the labour is performed by negro slaves. When the children express their outrage and willingness to forego sugar to alleviate the slaves' suffering, their mother responds by encouraging them in the fundamental Quaker spiritual practice of listening to the voice of God within. 'I admire the sensibility of your uncorrupted hearts, my dear children. It is the voice of nature and virtue. Listen to it on all occasion, and bring it home to your bosoms and your daily practice.'[6]

By the end of the nineteenth century, then, Quakers had a solid tradition of scientific practice, both in the workplace and as an acceptable pastime. They valued it for its spiritual insights and considered it to be inseparable from ethical concerns. Theologically, however, this was a time of turmoil for Quakers, as it was for Christians in general, as they sought to come to terms with new discoveries in science that showed that the Earth was not formed 6000 years ago (as calculated from genealogies in the Old Testament), and with the advent of 'higher criticism', which showed that the Bible was influenced by culture rather than being the infallible Word of God. As Rufus Jones, one of the Quakers we will consider in Chapter 2, would put it in strikingly visceral terms, 'It will be difficult, perhaps impossible, for my readers now living in peace in the lee of the dykes, to realize in any vivid way what it was like to be thrown into that open sea when the euroclydon was in full sweep.'[7]

Broadly speaking, two strands of Quakerism were apparent at this time. Evangelical Quakers wanted to keep literal interpretations of the Bible, for example, holding that creation unfolded as portrayed in Genesis. Liberal Quakers were convinced that religion should adapt to fit modern thought, accepted Darwinian evolution and higher criticism, emphasized that God was to be found within, and believed that theology had ultimately to be based on immediate experience of God (partly because appeals to the Bible and miracles were no longer credible).

A watershed event that saw the beginning of the dominance of the Liberal strand in Britain was the Manchester Conference of 1895. It was attended by over a thousand Quakers, and the burning issues of the day were made clear in an opening paper by Matilda Sturge: 'We live in an age of science, and its discoveries have brought new difficulties for those who have trusted in the verbal and literal inspiration of the Bible; we live in an age of criticism, and ... know that the Bible is literature rather than a book, that every part of it is not of equal value.'[8] It was thus necessary, she continued, to question how far the fundamental principles of Quakerism were fitted to the needs of the present day. As Henry Stanley Newman, a prominent English Quaker, put it in a letter to Rufus Jones in America, the Conference marked the first attempt of the Society to come to terms with modern thought and was of vital importance to stop young educated Quakers leaving the Society. Although there were contributions from both Evangelical and Liberal speakers, the Conference is generally accepted to mark the beginnings of the so-called Quaker Renaissance in Britain, characterized by an increasing engagement with modern thought and an emphasis on mysticism and social action. Its effects were felt in America, but Evangelical Quakerism remained (and still remains) stronger there than it did in Britain. The scientists we will meet in the next chapter, in both Britain and America, are representative of the Liberal tradition. There is a bit more information about the differences between Liberal and Evangelical Quakerism in the Glossary.

Of particular interest in relation to science is the session titled 'The Attitude of the Society of Friends towards Modern Thought', on the evening of 13 November.[9] In order to help us to gain some traction on the ideas that were being discussed, it is useful to look at four categories that are often used as ways of relating science and religion. They were proposed by Ian Barbour in the 1960s, as conflict, integration, independence and dialogue. The conflict category sees science and religion as almost at war, for example, setting literal interpretations of the six-day creation described in Genesis against theories of evolution. The independence category sees science and religion as having their own distinct fields of enquiry (e.g., science asks 'how' questions, and religion asks 'why' questions). The dialogue model sees science and religion as engaged in a conversation that leads to enhanced mutual understanding. Finally, the integration model resists the idea that the universe has sharply defined 'spiritual' and 'physical' components, insisting that explanations must encompass both.

So, returning to Manchester, in his introductory paper Thomas Hodgkin noted that the subject under discussion was 'The Relation of Quakerism to Modern Thought', clarifying that Quakerism was to be understood as 'that mode of apprehending Christianity in which the Society of Friends has differed from other Churches', and that 'modern thought' included 'scientism'. He made two points about science. First, he pointed out that George Fox and Quakers ever since have, under the influence of the Spirit of Christ, refused to call the Bible 'the Word of God'. This meant, he explained, that although the Bible contained many precious messages from God to the soul, the unscientific Hebrew chronology and cosmogony was not an essential part of Christ's message to the world today. Quakers can thus offer this approach to reading the Bible to those troubled by any clashes with science. Second, he claimed that although science may change the way we think about the universe, it would not change the way we thought about its Maker. Here, then, Hodgkin is illustrating Barbour's category of independence, because science will not affect the way we read the Bible or our view of God.

This attitude of independence was also apparent, albeit for different reasons, in the next paper. It was written by J. Bevan Braithwaite, the session's lone Evangelical contributor, but read in his absence by R. H. Thomas. Braithwaite advised Quakers to concentrate on practical holiness and, because life was short, to 'put a check upon many curious but unprofitable enquiries'. In other words, the ability to live a good life and form a holy character was, in Braithwaite's eyes, independent of science.

The biblical scholar J. Rendel Harris was, in contrast, adamant that science and religion could not be separated. 'This theory of the detachment of science and religion from one another never has been a working theory of the universe', he declared: 'the two areas must overlap and blend, or we are lost.' False science, for example, could lead to false doctrine, which Harris illustrated using a statement by the influential nineteenth-century Quaker reformer Elias Hicks, who claimed that God loved all people equally and placed them in the same condition as 'our first parents'. Modern man knows we never had any first parents, Harris objected, so insisting on this doctrine in opposition to science might cause people to doubt the rest of Hicks' statement, namely the love of God. He concluded by urging Quakers to embrace science, assuring them that suspicion of modern thought belonged 'more to the clergy than to a Society like our own'. Doing so, he confidently proclaimed, would bolster Quaker credibility, redeem the Society from the reproach of conservative timidity, and set Quakers in their rightful place among the intellectual forces of the world. Harris, then, is advocating dialogue as a way to make faith and doctrine fresh and credible.

The physicist Silvanus P. Thompson, whom we will meet

again in Chapter 2, followed Harris, explaining that he had been asked to consider the question of whether a scientific man could be a sincere Friend. The answer, unsurprisingly given his profession, was 'yes', and in fact he was confident that the practice of science and the plain speech favoured by Quakers reinforced one another, in that 'the habit of accurate thought and speech, of letting yea mean yea and no more, which is characteristic of Friends, is one that the scientific method tends ever to strengthen'. Echoing Harris, he too thought that modern thought would clear away only the human error that had grown up around divine truth. Crucially, though, science has its limits. Thompson insisted that 'To every man there comes a consciousness, not to be analysed in the test-tube of the chemist, nor probed with the scalpel of the physiologist, ... a consciousness of something quite other than those things which are to be apprehended by the physical senses'. So, Thompson sees the relationship between science and faith as multi-faceted. He is suggesting integration at the methodological level (how science is done), dialogue when it comes to doctrine, and independence when it comes to an experience of God.

In the final paper, the mathematician John William Graham stressed that the religious world now viewed the 'Indwelling Voice' as its central conception, a reference to the fact that Liberal Christian theologians more generally were emphasizing the importance of inner religious experience. Quakerism had always held this position, he proudly asserted, so held the future in the hollow of its hand. He too believed that Scripture is not infallible: even though in practice it excels all other books, its infallibility rests only on the ill-formed views of the bishops of the early centuries. Graham also touched on how God interacted with the world: 'In contemplating Divine Providence, modern religion regards it as constant, not as occasional ... The fatal result of claiming "Special" Providence is to banish God from the other ninety-nine per cent. of causation.' So here there is potentially yet another way of relating science and Quakerism, which Graham himself does not explore: God is involved in all physical processes, so perhaps science and religion should be integrated?

Thus all the Liberal speakers at the Manchester Conference embraced science, but in different ways. Views on the relationship between science and Quakerism encompassed independence, dialogue and integration, in various forms. The potential conflicts between science and religion were resolved on the one hand by giving authority to 'science', and on the other by claiming that Quakers had never viewed the Bible as infallible anyway.

The audience, however, were not all convinced that this was the right way forward. The recorded comments on the papers reveal a marked diversity of opinions, with one anonymous Friend objecting strenuously that 'I feel concerned to utter my earnest protest against the views uttered here to-night. It seems to me that this Conference, representing London Yearly Meeting, cannot do justice to itself without placing on record a protest'. Although the complaints are general rather than specific, it seems likely that the consternation felt by some centred on the approach to Scripture. Such was the strength of feeling that it was decided to delay further discussion until the next day, but when the appointed time came the clerk explained that to extend the discussion would interfere with the programme. He compromised by writing a minute that acknowledged the lack of opportunity for discussion, concluding that, 'We think it desirable that it should be distinctly understood, as applicable to all the sittings of the Conference, that it assumes no responsibility for the opinions expressed in any of the papers read before it.'

The Liberal agenda became dominant in Britain in the years to follow, and with it came a bullish attitude towards science. It is nicely shown in an open letter written in 1928 by Jesse Holmes, an American Quaker Professor of Philosophy and Religion, to the scientifically minded. It ends with an invitation: 'We have a faith, which we believe may properly be called a Christian faith, which has nothing to fear from science and which demands no medieval credulities of intelligent people. We invite such people to examine our faith and see if they do not belong with us.'[10] The next chapter thus looks at ten twentieth-century scientists or science teachers who certainly felt that Quakerism was where they belonged.

Further reading

Most of the material in the first part of this chapter comes from Geoffrey Cantor, *Quakers, Jews and Science* (Oxford Scholarship Online, 2005), some of which is summarized in his chapter 'Quakers and science' in *The Oxford Handbook of Quaker Studies* (OUP, 2015).

Chapter 2

Some twentieth-century Quaker scientists

Introduction

This chapter considers eight Quaker scientists and two Quaker reformers who taught science. They have been selected mainly on pragmatic grounds, as Quakers who happened to write about their faith and related it, to a greater or lesser extent, to their scientific practice. We'll consider briefly their scientific careers, their religious beliefs, and how they related science and Quakerism. We will meet them roughly chronologically, starting with Silvanus Thompson, a physicist born in 1851, and ending with Jocelyn Bell Burnell, a present-day astronomer.

Before meeting these scientists, however, it is useful to look briefly at two important Quaker concepts: pacifism and mysticism. Pacifism is important because most of the scientists considered were practising during a World War or the Cold War, and several took a stand for pacifism that reverberated through their careers. Mysticism is important because it is a notoriously vague term that, as we will see, keeps cropping up in relation to how these scientists talked about their faith.

The Quaker association with pacifism goes back to its earliest days, when Fox and a group of leading Quaker men issued a famous declaration to Charles II in 1660: 'All bloody principles and practices, we, as to our own particulars, do utterly deny, with all outward wars and strife and fightings with outward weapons, for any end or under any pretence whatsoever. And this is our testimony to the whole world.'[1] The declaration formed the basis of what became known as the Quaker Peace Testimony, and over the coming centuries Quakers wrestled with how to enact this in relation to various wars declared by governments. In particular, the work done by Quakers and other pacifist denominations established the recognition of religious pacifism, and conscientious objectors came to be categorized into three main groups: non-combatant (those willing to undertake service under the military in non-combat roles); alternative service (those willing to undertake service that was of social value in other organizations associated with the war effort); and absolute (those opposing any contribution to war).

The First World War, however, revealed the diversity of opinions held by Quakers on the Peace Testimony. Many took the conscientious objector route, and the official stance of the Society remained pacifist. Others, however, saw the matter as one of personal conviction and chose to serve in the military, for example, to protect their families or country. To address this issue, an international conference attended by more than a thousand Quakers was held in London in 1920. The conference stated that war could be eliminated by taking away its root causes, a task that required individual spiritual work and social change. This stance attracted many to Quakerism in the decades to come, including, as we will see, several of the scientists considered here.

Mysticism means different things to different people. The word *mystica* came into the English language through the fifth-century Syrian monk known as Pseudo-Dionysius the Areopagite, and derives from the Greek word *mu*, which has connotations of secrecy (as in 'keeping mum'). In the early eighteenth century, mysticism was not yet a recognized term, but mystical theology referred to a way of life involving prayer, contemplation and self-denial. By the mid-nineteenth century, the term mysticism was becoming more common, and interest in the subject was growing rapidly. Public and academic fascination was fuelled by the Transcendentalists in America (a movement associated with Ralph Waldo Emerson and Henry Thoreau, which held that the divine pervaded nature and humanity), the work of the psychologist William James on religious experience, and translations of works from different religions, which suggested that religious experience transcended religious boundaries.

In the early twentieth century, popular books on mysticism included *The Mystical Element in Religion* (1908) by the Catholic Friedrich von Hügel, *Mysticism* (1911) by the Anglican Evelyn Underhill, and *Studies in Mystical Religion* (1909) by the Quaker Rufus Jones. The three authors knew and admired each other, but held different views on what mysticism involved and what constituted a mystic. It is Jones' view that is most relevant here. Crucially, he insisted that mysticism wasn't esoteric or reserved for a few ascetics. Rather, it was a natural, everyday experience of a personal God that led to action in the world and was, to a greater or lesser extent, open to everyone. Furthermore, he claimed that Quakerism was a mystical religion because it was based on this first-hand experience of God. The Quaker scientists considered in this book may, if pressed, have given different definitions of mysticism, but they are likely to have been influenced by Jones and to have in mind not ecstasy but an undramatic experience of God that could involve a feeling of peace or a sense of how to act in a certain situation.

Further reading

There is more information on the Quaker attitude to war in the chapter by Lonnie Valentine, 'Quakers, war, and peacemaking', in *The Oxford Handbook of Quaker Studies* (OUP, 2015). Mysticism is explored in the Quaker Quicks book by Jennifer Kavanagh, *Practical Mystics* (Christian Alternative, 2019).

2.1

Silvanus Phillips Thompson: The quest for truth

Silvanus Phillips Thompson (1851–1916) was a prominent physicist and much respected Quaker elder. His scientific interests ranged from electricity and magnetism to optics, but he is remembered more for his teaching than for new and notable discoveries. He was widely acclaimed for his Friday Evening Discourses at the Royal Institution and for his Christmas lectures to children, and his charming 1910 textbook *Calculus Made Easy* was written in a conversational style and is still in print in updated form. (The book sought to cure engineers of their fears of this reputedly difficult mathematical technique, with Chapter 1 titled 'To deliver you from the preliminary terrors'.) As a preacher, he was widely admired for the breadth of his knowledge, his ability to weave together similes from history, science, art and literature, and the courage, fervour and vision of his addresses.

Thompson was born into a cultured Yorkshire Quaker family. His mother was a botanist and a talented botanical illustrator. She passed her artistic talent on to her son, who was known for his quirky caricatures and, later in life, his watercolours, which he sometimes exhibited with the Royal Water Colour Society. His father was a teacher at Bootham, which we came across in the last chapter, and Thompson started as a pupil there in 1858. After school, he studied at the Quaker Training College in Pontefract while simultaneously preparing for a BA in classical studies at the University of London (which at the time was purely an examining body). After graduating, he returned to Bootham as a junior master, but was frustrated by the lack of scientific equipment and soon enrolled as a full-time student at the Royal School of Mines in South Kensington, studying chemistry and physics. In 1876 he was offered a lectureship in physics at the recently established University College at Bristol, and in 1878, at the age of 27, was appointed to the Chair of Physics. In 1885 he became Principal and Professor of Physics at Finsbury Technical College, staying there for 30 years.

The search for truth was central to Thompson's personal and professional life. We saw in the last chapter that he was one of the speakers at the 1895 Manchester Conference, where he had enthused that science would clear away the misconceptions that had grown up around religious doctrine. In 1915 he gave the Swarthmore lecture (a Quaker annual series of lectures that has run from 1908 until the present day), published as *The Quest for Truth*, exploring what this quest means in science and religion. It requires continued effort and courage, he advised, and is something every person must do for themselves. It also involves being comfortable with doubt: 'The craving for certitude is not in all respects a sign of spiritual health', he warns. In fact, 'the very eagerness to be certain tends to vitiate the search by a temper of impatience'.[2] Using a scientific analogy, he explains that at times it is even necessary temporarily to hold two propositions that are mutually incompatible, until more data become available.

At the time of the Swarthmore lecture, Thompson was also working on a book, *A Not Impossible Religion*, which was published after his death in 1918. The introduction reveals something of the challenges he faced in his own quest. At one point he moved away from Quakerism briefly towards Anglicanism, he says, where the 'pretensions of clerics and their bigotry towards science' drove him towards agnosticism.[3] Nevertheless, he continued to study Scripture and other religious writings and came slowly, and with twenty years of heart-searching, to the convictions in the book.

One of the results of this long quest was that Thompson was

very sure about what he did not believe. Alluding to the fact that the death of Jesus has often been interpreted in terms of older Hebrew ideas about the necessity of a blood sacrifice to reconcile humans and God, he complains that creeds arose out of the primitive beliefs and mythologies of an Oriental people. To impose them on the Western, scientifically enlightened mind, he says, is 'an outrage on the spirit of truth and an insult to the soul of man'.[4] Indeed, the very idea that God would require a blood sacrifice is monstrous. What he did believe was that Christ claimed a special Sonship and a Divine mission, but not identity or equality with God, and that the resurrection was central to the life of the disciples because it transformed them from shame-faced followers ready to deny their faith to indomitable leaders and evangelists. Furthermore, salvation is not about life after death but is for this life. It is freedom from sinning, not from the consequences of sin (the position taught in many churches), because Christ has given his followers a new life of righteousness.

When it came to the relationship between science and religion, Thompson exhibited some of the attitudes we saw in the last chapter. He was convinced, for example, that science should be practised alongside social responsibility, and in fact was known for peppering his physics lectures with exhortatory phrases such as 'do the duty that lies nearest you'.[5] Furthermore, just as we saw that some earlier Quaker scientists had worried that an excessive interest in science might be detrimental to their spiritual life, Thompson warned that although openings towards a career of greater prominence came from time to time, it was necessary to pray for guidance in case the greater responsibility would be a distraction from the primary duty of simple service. He personally seemed to feel that he had an obligation to devote some of his ability and influence towards ameliorating the conditions of the working classes, and this may have been one reason he championed the cause of technical education—he devoted several vacations to studying the provisions for technical education in France, Germany and Switzerland before taking up the position at Finsbury.

A further way in which Thompson saw science and religion as being related was through intuition. In *A Not Impossible Religion,* he points out that just as poets, musicians and artists work with a sort of instinct, so do scientists. He had written an acclaimed biography of Michael Faraday, the English physicist who did much to unravel the mysteries of electromagnetism, and explains that Faraday had an extraordinary faculty of vision that did not employ reason but arrived at rational results by another process. By being familiar with experimental facts and letting his thoughts play freely with them, Faraday was able to see new facts and relations which were subsequently verified by experiment. This quality of intuition, Thompson felt, is also apparent in spiritual discernment, when reasoning can no longer find the way ahead.

As a final example, Thompson suggested that the existence of laws in the 'outward' universe has a parallel with the existence of laws in the spiritual world, knowledge of which can be gained by a process similar to that used in science, namely observation, generalization, induction (inferring a general law from particular instances), and analysis. The 'law of health', for example, states that just as in the natural world a healthy life is one of alternating activity and rest, so in the spiritual world the spiritual faculties must be used and exercised but also permitted times of repose. The 'law of growth', which reflects Thompson's continual striving after truth, is, he warns, broken by the assumption that the revelation of God to man ended with the last word of the Bible.

Thompson thus comes across as being a wise scientist with a well-developed social conscience whose career decisions were made in accordance with his Quaker faith. For him, both science and religion involved searching for truth, accepting doubt, and exercising intuition and integrity, and they both revealed testable laws.

Further reading

Both of Thompson's books, Silvanus P. Thompson, *The Quest for Truth* (Headley Brothers, 1915) and Silvanus P. Thompson, *A Not Impossible Religion* (William Clowes and Sons, 1918), are available online through archive.org.

2.2

Sir Arthur Stanley Eddington: Seeking in science and religion

Arthur Eddington (1882–1944) was born into a Quaker family in Kendal. After school, he studied at Owens College, part of the University of Manchester, before winning a scholarship in Natural Science to Trinity College, Cambridge. In 1906 he started work at the Royal Greenwich Observatory, and in 1913 was elected Plumian Professor of Astronomy and Experimental Philosophy at Cambridge. He is known for his work on the structure of stars (readers with an interest in astronomy may well be familiar with the Eddington limit, which describes the maximum luminosity of a stable star), and for his experimental verification and popular exposition of Einstein's theory of relativity. Eddington never married, and his sister, Winifred, fulfilled the social role of professor's wife until his death. He had a lifelong passion for cycling, displaying his mathematical turn of mind by tracking his progress with an '*n*-number', where *n* was the number of times he had cycled more than *n* miles.

Eddington was a student at the start of the Quaker Renaissance instigated by the Manchester Conference. He was heavily influenced by the mathematician John W. Graham, who was the principal of Dalton Hall, the Quaker residence where Eddington stayed while studying at Owens. Graham was a prominent Liberal Quaker leader and, as we saw in Chapter 1, had been one of the speakers at Manchester. He presented Quakerism as a religion based on mysticism, namely on an experiential knowledge of God. Both mysticism (as opposed to a religion based on fixed doctrine) and science, he insisted, involved a search for truth, which would in turn benefit society. Eddington was also impressed by Rufus Jones (whom we will meet next), who gave a talk on mysticism at the Kendal summer school that Eddington attended in 1908.

In addition to the momentous developments that were occurring in Quakerism in the early twentieth century, there were also tremendously exciting and revolutionary developments in physics, as countless curious and counterintuitive discoveries were being made about the fundamental properties of matter and the nature of space-time. These discoveries seemed to have compelled Eddington to develop his philosophy to a greater extent than did the other Quaker scientists we consider here: 'In approaching the more fundamental problems of science the physicist finds it necessary to adopt a more philosophical outlook than is habitual to him', he wrote in a 1933 article titled 'Physics and philosophy'. His philosophy will not be discussed here, but essentially he was a philosophical idealist, believing that mind or consciousness is the most direct thing in our experience, and that everything else is inferred from that. There is a certain amount of correspondence between the physical world and what we experience because of natural selection, he said, but ultimately the physical world can be reduced to the readings of scientific instruments, and the fundamental laws and constants of physics are wholly subjective, a creation of mind.

Eddington held that, far from developments in physics threatening religion, they in fact made it more credible. In his 1929 Swarthmore lecture, for example, he explains that electrons are no longer thought of in terms of concrete billiard balls but in terms of symbols and mathematical equations. The fact that matter is reduced to a shadowy symbolism, he says, has the consequence that 'we are no longer tempted to condemn the spiritual aspects of our nature as illusory because of their lack of concreteness'.[6]

Eddington also saw a number of parallels between religion and science. Echoing Graham, he was convinced that they are both based on experience: 'If science claims in any way to be a guide to life it is because it deals with experience, or part of experience,' he said. 'And if religion is not an attitude towards experience, if it is just a creed postulating an ineffable being who has no contact with ourselves, it is not the kind of religion which our Society stands for.'[7] In fact, he maintained, the sense of a divine presence irradiating the soul is one of the most obvious things of experience to some.

Also echoing Graham, both science and religion involve seeking. If we were told that the mysteries of the universe would be solved in a few years, the tidings would be by no means joyful. On the contrary, 'In science as in religion the truth shines ahead as a beacon showing us the path; we do not ask to attain it; it is better far that we be permitted to seek.'[8] In Quakerism, seeking involves paying attention to the Inward Light (which is discussed in a bit more detail in the Glossary) for new revelations of God rather than relying on creeds. In physics, it means that the products of science should be judged not by their merits but for how well they could support further seeking: science, Eddington felt, should be viewed as an engine, not a building. In practice, this meant that when he calculated the temperature of stars, for example, he warned that his approach was more utilitarian than rigorous. It was an approach that his fellow astronomer James Jeans derided as involving unforgiveable sloppiness, and their opposing views led to a series of lively and competitive debates in the halls of the Royal Astronomical Society that would become legendary.

A further way in which science and religion were interwoven in Eddington's life is that his Quaker values affected the actual scientific research that he pursued. Graham had argued that the corollary of pacifism in the world was internationalism, in that the best way to prevent war was to work against the nationalism and historical divisions that drove conflict. During the First World War, many British scientists and organizations were breaking contact with their German counterparts, but Eddington argued that science was an enterprise that was inherently international. Einstein was, of course, a prominent German-born scientist (although he surrendered his German citizenship when he moved to Switzerland in 1895), and, at a time when his work was being shunned by many British scientists, Eddington became the chief exponent of Einstein's theory of relativity. It was known that in 1919 an eclipse would be visible from some parts of the world, making it possible to verify the theory by measuring the minute deflection of starlight that would be apparent when the moon obscured the light of the Sun. Eddington managed to secure funding to travel to Principe, an island off the west coast of Africa, to make observations. It was a gruelling expedition, involving difficulties in transporting bulky equipment in the aftermath of war, bad weather, mosquitoes, and marauding monkeys. Although there was, naturally enough, some debate over the interpretation of the photographic plates that Eddington obtained, most astronomers agreed that a deflection of starlight had occurred, lending credence, if not absolute proof, to Einstein's theory.

Eddington, then, managed to weave together physics, philosophy and his experiential Quaker faith. And the practice and values of that faith affected both the way he did science and the science that he did.

Einstein, incidentally, later expressed his appreciation of Quakers. He gave a lecture at Swarthmore College, Pennsylvania in 1938, not realising that it was a Quaker institution until told so by his translator. In a hastily written commendation scribbled on the back of a used piece of paper he wrote that 'With admiration and respect I have seen, in the course of many years, how successfully and selflessly the Religious Society of Friends worked in the entire world to lessen human suffering and to make the teachings of Christ apply to real life'.[9] The note made its way to Howard Brinton (whom we will meet later), who had trained as a physicist. Unsurprisingly, he was fascinated by Einstein's equations on the other side of the paper, and attempted to work through them with one of his students.

Further reading

A very readable account of Eddington's life and thought is Matthew Stanley, *Practical Mystic* (University of Chicago Press, 2007).

Eddington's involvement with Einstein is the subject of a 2008 BBC film titled *Einstein and Eddington*.

2.3

Rufus Jones: Psychological light on the Inner Light

Born in a farmhouse in rural Maine, Rufus Jones (1865–1948) is widely acknowledged to be a Quaker giant of the twentieth century. His childhood is vividly recounted in a couple of his autobiographical books, where we read of outdoor adventures and misadventures with the local farm boys, magnificent forests and glorious sunsets. He describes daily times of family silent worship in which he sensed gleams of eternal reality breaking through, and a religious culture that welcomed visiting Quakers as messengers with words from God. Jones spent much of his career teaching philosophy and psychology at the Quaker Haverford College in Philadelphia, but his influence reached far beyond the confines of the campus. He vigorously promoted the Liberal agenda within Quakerism, and is particularly well known for arguing that mysticism is a felt relationship with God available to everyone. An awareness of God is not esoteric but a normal part of everyday life, he insisted again and again, and this experience should equip the mystic (i.e., you and me) to make the world a better place. His readable books on mysticism and the spiritual life were both numerous and popular, not just among Quakers, and helped to popularize mysticism in the first half of the twentieth century. When America joined the war in 1917, Jones put his convictions into practice and was instrumental in setting up the American Friends Service Committee. The organization gave American Quakers the chance to serve in non-combatant roles and later to carry out relief work, and Jones acted as chair on and off for the rest of his life. In 1947, along with the British Friends Service Council, it was awarded the Nobel Peace Prize.

Jones argued frequently and eloquently that Quakerism needed to be brought in line with the latest developments in science, an attitude that was fundamentally shaped by his school science teacher, Thomas J. Battey. Darwin's *On the Origin of Species* had been published in 1859, and Battey had studied at Harvard with the well-known naturalist and Darwinian Asa Gray. It was in Battey's class that Jones first heard the 'astonishing fact' that the world was not made in six days 6000 years ago, but fortunately Battey offered an interpretative framework that would inspire Jones for the rest of his life: he 'carried us over from our childish idea of a God who worked from the outside like a mechanic to the higher conception of a God who works from within as a living creative energy. He helped us to realize that the account in Genesis is a great poetic story ... I leaped forward to the new view and with it I won my spiritual freedom'.[10]

Jones also told of another breakthrough that occurred while he was an undergraduate at Haverford and presumably still grappling with the questions raised by science. It was on the day his teacher Pliny Chase brought a thick new book to class, claiming it to be by a prophet of the new age. The book was *Natural Law in the Spiritual World*; the prophet was the Scottish biologist Henry Drummond. Drummond argued that natural laws corresponded to spiritual laws, and drew spiritual analogies from natural phenomena ranging from parasites to the environment. Jones enthused that, while the book was imperfect, it came 'like water to shipwrecked men'.[11]

These attitudes to science formed at school and college prepared Jones well for the Liberal approach of the Manchester Conference, and he wrote about science and religion in many of his books. *Social Law in the Spiritual World* (1904), for example, took its title and inspiration from Drummond. His aim, he said, was to do for psychological laws what Drummond had done for natural ones, namely to show how they shed light on the relationship between God and humans. As he put it in the Introduction, when Drummond wrote his book the prevailing problems for Christians seeking to reconcile science and faith came from biology, but now they came from psychology, as every precious article of faith must submit to a psychological test. Can Christians bring their ship past this new headland, he asked? Yes, he assured his readers, and in fact the cure for scepticism is always deeper knowledge. Essentially, Jones held that faith is not endangered by the advance of science but by the stagnation of religious concepts. If religion halts at some primitive level and science marches on to new conquests then of course there will be difficulties. Thus while our hearts would never be satisfied by a God who could be found by science, it is at the same time true that the progress of science can greatly clarify our ideas about the kind of God we have a right to expect to find.

Jones was particularly interested in the work of the so-called father of modern psychology, William James. James lectured at Harvard at a time when psychology was just beginning to emerge as a scientific discipline, an event that is usually taken to begin with the opening of the first scientific laboratory in Leipzig, Germany, in 1879. In the years to follow, psychology, suffering from what has been termed 'physics envy', sought to beef up its scientific credentials by establishing labs, journals and societies. At the same time, however, many psychologists were fascinated by psychic phenomena such as telepathy, spiritualism, and mediums.

James straddled these two faces of psychology, adding a Ouija board to his laboratory equipment and spending many years investigating the celebrated Boston medium Leonora Piper. His seminal textbook *Principles of Psychology*, published in 1890, ran to 1400 pages, had been 12 years in gestation, and instantly became a classic that has influenced generations of psychologists. In *The Varieties of Religious Experience*, published in 1902, he gathered together numerous case studies of religious experience. In the final chapter he mused about a 'more' beyond our everyday experience with which we feel some connection. James refused to associate the 'more' with any religious tradition, or, as he put it, set of 'over-beliefs', so there is no reason why the 'more' should be the Christian God, for example. He claimed merely that 'whatever it may be on its *farther* side, the "more" with which in religious experience we feel ourselves connected is on its *hither* side the subconscious continuation of our conscious life'.[12]

Jones, meanwhile, had already begun to suspect that mysticism was best studied through psychology and philosophy, so when he came upon *Principles* in a local library he was immediately smitten, exclaiming in his autobiography that 'no one with my interests could forget an event like that!'[13] He wrote to James, and the two corresponded for many years. He also used James' books when teaching psychology, later reminiscing about the 'solemn awe' that came over a class when he introduced them to James' chapter in *Principles* about how to create good habits.

Most significant, however, is how Jones used James' idea of the 'more' to reformulate the traditional Quaker understanding of 'that of God', or the 'Inward Light' in a person. Traditionally, the divine Inward Light was viewed as being separate from humans, like a candle in a lantern. Now, Jones proclaimed, psychology had shown that there was no separation between humans and God. Thus the Inner Light (Jones' preferred terminology) was an inherent part of human nature. Essentially, Jones was drawing on James to argue that the subconscious was where God met with humans: it was a 'Shekinah of the Soul'. Whereas James had been agnostic about the nature of the 'more', however, Jones associated it with the Christian God of love. Furthermore, he was undoubtedly cherry-picking when it came to his choice of what psychological research to draw on. Other psychologists, such as James Leuba, for example, were arguing that psychology had shown that God did not exist (a discovery that would initially cause despair but would eventually be for the best, he felt).

Evangelical Quakers were quick to criticize Jones, complaining that his views contradicted Scripture, in that a relationship with God was possible only through Jesus. Liberal Quakers gradually accepted his ideas though, and part of Jones' legacy is held to be his new formulation of the Inner Light. In the 1950s, Liberal Quakers began to question the need for their faith to be tied to its historical Christian roots, and now, in the twenty-first century, many Liberal Quakers identify as 'non-theist' (i.e., as not believing in God). There have been a number of factors at play in these shifts, but it is interesting that these positions are consistent with William James' view of the 'more'. They are, though, perhaps not developments that Jones would personally have agreed with, given that he himself strongly self-identified as a Christian.

Jones was a charismatic figure, often described as radiant. He was also staggeringly productive, as judged by his 50-plus books, teaching, lecturing, involvement with Quaker affairs, and work with the American Friends Service Committee. His attempts to bring Quakerism in line with psychology have had long-lasting effects on Quaker faith and practice.

Further reading

Jones' biography is Elizabeth Vining, *Friend of Life* (Forgotten Books, 1958).

A number of Jones' books are available online through archive.org.

2.4

Edwin Diller Starbuck: Psychology in the service of religion

Edwin Starbuck (1866–1947) was brought up on a farm outside Indianapolis. The rural lifestyle in the midst of a network of small Quaker villages was very similar to that of Jones, as Starbuck recalls daily family times of Bible reading and silence, weekly rhythms of Quaker meetings, and a close-knit community in which neighbours vied with each other for the privilege of taking the night shifts when caring for the sick. After graduating from Indiana University, he taught mathematics at Vincennes University, but became fascinated by Max Müller's book *Lectures on the Science of Religion*. As a result, he decided to pursue his interests in the psychology of religion at Harvard in 1893–1895. It was here that he met his future wife, Anna Maria Diller, later taking Diller as his middle name. After a series of short-term positions at various universities, he spent the years 1906–1930 at the University of Iowa, before moving to the University of Southern California for the remainder of his career. His calling, he said, was to 'try to render thinkable and usable the illusive reals of religion'.[14]

Starbuck wrote an autobiographical essay 'Religion's use of me' in 1937. He comes across as a fiercely intelligent and complex character, describing how opposing sides of his personality (tenderness and toughness, acceptance and doubt, sensitivity and vigorous intellectuality) had many a 'jolly tussle' as well as some long, drawn-out conflicts. He experienced struggles and sleepless nights over the challenges to the authority of the Bible and the battle between Adam and the monkey, as he put it, but eventually found a lifeline when he realized that instead of trying to *make* truth through thinking he had to *find* truth, accept it, and take the consequences. He described himself as having a persistent ineradicable mysticism, which was accompanied by a matching theology that was a sort of pantheism, or a sense of an Interfusing Presence, and attributed his tastes, attitudes, temperamental eccentricities and religious peculiarities to his Quaker upbringing.

Starbuck had a fascination with conversion, which seems to have originated in an incident in his youth. Revivalism was sweeping through America in the late nineteenth century, and although the idea that one must have a definite conversion experience was in many ways contrary to Quaker beliefs, many Quakers were caught up in revival meetings. Starbuck himself stood up and made a confession of faith at one, only later realising that this implied that he had not been devoted to God previously, and concluding that he had been emotionally manipulated. He recalled that his older brother Elwood also 'fell for the same stunts', and went into seclusion for days afterwards, feeling that he had made a fool of himself.[15] In spite of these reservations, Starbuck recognized that there was some spiritual vitality in the experience, and set out to separate this vitality from what he deemed its unseemly and unnatural context, thus recovering its power, he hoped, for the modern believer.

His guiding principle in exploring conversion, which has an unmistakeably Quaker flavour to it, was that he wanted to deal primarily with the first-hand religious experience of individuals, and not with their theories about religion. To this end, he devised and circulated questionnaires on conversion to friends and churches. The questions were deeply personal. Starbuck started by asking 'What errors and struggles have you had with (a) lying and other dishonesty, (b) wrong appetites for foods and drinks, (c) vita sexualis', before moving on to ask about the feelings and experience of conversion and its aftermath in relation to God, nature and people.[16] The religious reformer Thomas Wentworth Higginson protested that the questionnaires amounted to moral and spiritual vivisection, but enough people responded for Starbuck to be able to compile numerous charts and graphs measuring doubt, anxiety and intensity of experience. He found laws and patterns, such as conversion being a distinctively adolescent phenomenon, and more likely to occur in people who were dominated by feelings than in those who were 'intellect predominant'.

William James was initially sceptical about Starbuck's use of questionnaires, but later changed his mind, and in fact drew heavily on Starbuck's reports in his immensely influential 1902 book *The Varieties of Religious Experience*. In the preface to Starbuck's *The Psychology of Religion*, James wrote that Starbuck had brought conciliation into the long-standing feud between science and religion: whereas evangelical extremists saw conversion as a supernatural event, and a scientist saw it as nothing but hysterics and emotionalism, Starbuck had shown that it was often a perfectly normal psychological crisis that was merely regulated by the evangelical machinery.

As these endeavours illustrate, Starbuck's goal was to reveal the laws and processes at work in the spiritual life. These spiritual laws, he thought, would lead to greater wisdom in religious education, increase our appreciation of spiritual matters, and lift religion out of the domain of feeling so it could appeal to the understanding. In turning this desire into practice, he invested a huge amount of energy into the religious education of children and into seeking the best way to meet the religious needs of young men. Regarding children, he abhorred the idea of indoctrination and argued that instead the integrity of children's personalities should be respected by intriguing their imagination and eliciting their creative interests. Regarding young men, he similarly argued against programmes of convincement, instead advocating a programme of enrichment that included, for example, finding spiritual values in recreation.

Starbuck was also fascinated by the role of feelings and bodily sensations in religion. For a long time he struggled with Cartesian dualism, the idea that body and mind are separate, but in a tantalizingly brief description of his resolution to this problem he tells of a sudden 'metaphysical illumination' on a walk, when he realized that 'I, a mind, a body-mind, am in and of a universe of meaning'.[17] Given his observation about the lifelong effects of his Quaker upbringing, it could well be that childhood memories of his father influenced him here. Speaking out loud against the backdrop of silence in a Quaker meeting involved an inner battle for many Quakers, as they sought to discern whether the message they felt compelled to share was from God or not. Starbuck recalls that his father was a gentle, modest man and would have much preferred that the Spirit would leave him alone. When in the midst of this inner battle, his face would colour a deep red, and in spite of himself his lips would begin to move and his body to show signs of restlessness, 'and not infrequently did the Spirit win against the reluctant flesh'.[18] The non-Cartesian interplay of emotions, thoughts, bodily sensations and spiritual inspiration could hardly be clearer.

Starbuck explored the physical side of this mind-body link at the University of Zurich in 1903–1904, where he learnt how to use the sphygmomanometer to measure blood pressure, and other devices to register the relationship between bodily changes and mental processes. Furthermore, his conviction that the mind-body exists in and is of a universe of meaning is apparent in his argument that religious feelings point to the existence of God: because they are universal, and because most other inner dispositions can be relied on for true knowledge of the world, these feelings must point to the existence of an objective spiritual reality.

Starbuck's mission and optimistic hopes for psychology are

encapsulated towards the end of his essay: 'If the promises of the last half-century are fairly realized there can be no doubt that we shall be moving speedily with the aid of clear-headed cultural engineering toward the eradication of much moral ugliness and spiritual failure and toward the realization of a better sort of humanity.'[19]

Further reading

Starbuck's autobiographical essay is 'Religion's use of me', in *Religion in Transition*, ed. by V. Ferm, pp. 201–260 (George Allen & Unwin, 1937). It is available online through archive.org.

2.5

Howard Brinton: Quakerism as a method

Howard Haines Brinton (1884–1973) was born in West Chester, Pennsylvania, and brought up as a 'Wilburite' Quaker, namely in a branch of Quakerism that emphasized particularly strongly the importance of relying on the Inward Light rather than on the Bible or human initiative. He studied under Rufus Jones at Haverford, receiving a BA in 1904 majoring in maths and physics, and stayed on for a further year to study for an MA. His teaching career got off to an inauspicious start at Friends Select School in Philadelphia: he was hired to teach maths but his inability to discipline the pupils led to an early firing. ('John Woolman says in one of his essays that a teacher may appeal to that of God in each pupil', Brinton wrote. 'Apparently, as far as my experience went, such an appeal is ineffective.'[20]) His next appointment, at Olney Friends School in Barnesville, Ohio, was happier and more successful, but he left in 1908 to pursue a second MA at Harvard, where he studied under William James, among others. From 1909 to 1934 he held a variety of roles, including teaching maths, physics and religion at a number of colleges. While teaching physics at Earlham College he studied for a PhD—not in physics, but on the sixteenth-century German mystic and philosopher Jacob Boehme, obtaining his doctorate in 1925. Jones had introduced Brinton to Boehme, and he had become fascinated with Boehme's efforts to reconcile the inner world of mysticism with the outer world of science (as it was understood at the time). From 1934 to 1952 he held the influential post of director and theology lecturer at Pendle Hill, a centre set up in 1930 near Philadelphia to equip Quakers to serve in the wider world.

Brinton published many pamphlets and a few books on

Quakerism, some of which have had a significant effect on how Quakerism is understood. They contain a number of unusual and interesting parallels between Quakerism and science, so we'll look at a few of them here.

He argued, for example, that there were similarities between Quakerism and the scientific method. Quakerism is primarily a method, just as science is primarily a method, he wrote in the Introduction to his 1952 *Friends for 300 Years*, a classic that is still available in an updated edition. The aim of the book, he wrote, was not to produce a history of Quakerism but by means of historical illustration to examine a method. Protestant, Catholic, and Quaker practices can be identified with the lecture, lecture demonstration and laboratory methods, he explains. Acknowledging that he is simplifying things, he points out that Protestant worship is centred in the sermon, Scripture and hymns, which is equivalent to the lecture method, in which the teacher expounds in words the reality that is being presented. In Catholic worship, the priest not only speaks of the divine but reveals the divine to the congregation in the Sacrament, a practice equivalent to the lecture-demonstration method, in which the teacher presents the facts by performing experiments in which the facts are illustrated. In a Quaker meeting, however, an opportunity is offered for each individual to practice the presence of God as an experience of their own, which is equivalent to a student conducting experiments in a laboratory.

Furthermore, in his *Guide to Quaker Practice*, he points out that the method of Quakerism and the scientific method are both based on some definite belief. The scientist's method is based on the theory that the universe is a cosmos, not a chaos, and so under similar conditions the same causes will produce the same results. The Quaker method is based on belief in a God-centred spiritual universe, the inner truth and meaning of which are to some extent accessible to humankind. It is also the case that both methods take past discoveries into consideration:

in Quakerism this is the accumulated wisdom of the saints and prophets that have gone before.

Science has its limits though, and Quakerism can complement it. For an individual, for example, a Quaker meeting provides an opportunity to centre down and focus on 'those profound regions where the ultimate meaning of life is discovered'.[21] This practice is particularly important for intellectuals, who can be 'one-sided' because of their tendency to rationalise and analyse. Furthermore, scientific understanding can only go so far in explaining human life. The problem with the attempt to explain life in terms of physical and chemical reactions, he says, is that no one really believes that life is a mechanism—we don't treat people as complicated machines, and we have a sense of freedom.

One of Brinton's major contributions, arising from his belief that Quakerism is a method, is that in the *Guide* he distilled for the first time the so-called Quaker 'testimonies'. These testimonies are today so closely associated with Quakerism that many do not realize that they were not named as such until 1943. In her book *Testimonies,* Rachel Muers explains that testimonies are patterns of action and behaviour that are understood as an individual and collective response to God's leading and call. They are shared, intergenerationally sustained, develop over time, and are to be practised in everyday life. The precise list has evolved slightly since Brinton, and today they are often referred to by the acronym SPICES: simplicity, peace, integrity, community, equality and sustainability. Brinton had 'discovered' these testimonies in the same way as a scientist 'discovers' a law, writes his biographer. He surveyed a jumble of Quaker writings and from them distilled four social testimonies, namely simplicity, harmony, community, and equality, and one personal testimony, integrity.[22] While this simple list has undoubtedly eased the task of explaining Quakerism to non-Quakers, some have complained that it reduces spiritually gained insights to a list of behaviours.

A further preoccupation of Brinton was related to new ideas that were being put forward in the early twentieth century in the face of Darwinism, namely a worldview in which God's creative power was at work through nature. A number of scientists were advancing a vision of nature in which progressive, purposeful evolution led up to the human mind. Ideas and terms that were circulating included the 'becoming' and 'elan vital' of the French philosopher Henri Bergson; the 'Nisus', or striving of all nature to God, of the British philosopher Samuel Alexander; and the 'holism' of the South African statesmen and philosopher Jan Christian Smuts, which described the movement towards wholeness in the universe. It was an approach that was plausible in the early decades of the century, but became increasingly less relevant for science as the theory of natural selection was revived through a synthesis with genetics in the 1920s and 1930s.

Brinton developed these ideas and linked them with the Inner Light, enthusiastically citing all the above thinkers in his 1931 Swarthmore lecture *Creative Worship*. In one particularly poetic passage he traces the evolution of the universe, describing how the initial swarm of electrons formed atoms under the Power which unites and uttered 'the creative *Fiat*', how the atoms then 'jostled and fought until again the Spirit of Cooperation entered and they combined to create molecules', and so on until 'the infinitely elaborate structure of a human brain'.[23] Significantly, he links this creative process with what happens in Quaker worship, which is not based on a traditional creed or sermon but, like the universe itself, is open to the creativity of God: 'Worship based on the Inner Light which is also the Inner Life', he affirms, 'is as open to the novel and unexpected as is life itself.'[24]

Nearly 40 years later, in his pamphlet *Evolution and the Inward Light: Where Science and Religion Meet*, Brinton returned to these ideas. A Friends meeting feels that it is being pulled up by a Divine power beyond it and above it, he explains, and we can think of the primordial Logos, or the Word of God in the prologue of John's gospel, as generating a field of spiritual force to gradually pull our world toward itself into a single unity. Readers who are familiar with the work of the Catholic priest and mystic Pierre Teilhard de Chardin may well recognize resonances with his work (which was initially banned and not published until after his death in the mid-1950s), but Brinton's approach is distinctively Quaker, identifying as it does the Logos with the Inner Light.

Brinton was thus a hugely influential Quaker whose contributions to Quakerism's self-understanding owed much to how he thought as a physicist and his interest in mysticism.

Further reading

Brinton's biography is Anthony Manousos, *Howard and Anna Brinton: Re-inventors of Quakerism in the Twentieth Century* (Quakerbridge, 2013).

Brinton's book is *Friends for 300 Years* (Pendle Hill Publications, [1952] 1994). Some of his other works are listed in the bibliography.

2.6

Lewis Fry Richardson: The statistics of deadly quarrels

Born into a prosperous Quaker family in Newcastle-upon-Tyne, Lewis Fry Richardson (1881–1953) was another beneficiary of the extraordinary Bootham School, where he boarded from age 12. There, he said, he caught glimpses of the marvels of science from James Edmund Clark, and was struck by the insistence of Alfred Neave Brayshaw that science ought to be subservient to morals. Richardson excelled in the school's Natural History Society, winning prizes for his natural history diary in 1897 and 1898, and also in 1898 the Bootham natural history scholarship. (His projects had included displaying an insect collection of 167 species, making plaster casts of animals and birds, and studying bacteria in putrefying solutions.) He went on to study at Durham College of Science in Newcastle for two years before going to King's College, Cambridge, leaving in 1903 with a First in Natural Science. During his career he published numerous papers on topics including meteorology, war causation, scientific instruments, and psychology.

Richardson held a variety of posts throughout his life, and once mused about how his love of solitude manifested itself in his work: 'When solitary, I am usually serene; when in a crowd, I am often embarrassed ... After I had become a professional scientist and when I met obvious leaders, my usual tendency was to shy away from them.'[25] Some jobs, for example, at University College Aberystwyth and Manchester College of Technology, involved teaching, but Richardson found it difficult to maintain discipline. Inevitably it was a difficulty that was ruthlessly exploited by his pupils, and on one occasion he ruefully recalled being unable to bring order to a class who took to humming and making farmyard noises. In 1913 he was appointed as Superintendent at Eskdalemuir Observatory in Dumfriesshire, where he was responsible for making weather and geophysical observations. When war broke out, the nature of his work meant that he could have stayed on, but instead he resigned and joined the Friends Ambulance Unit in France, where he served between 1916 and 1919. He was torn between an intense curiosity to see war at close quarters, he admitted, and an intense objection to killing people, with both desires mixed with Quaker-inspired ideas about public duty and doubt as to whether he could endure danger. He continued his research while working as an ambulance driver, and a colleague recalled how Richardson used to set up meteorological instruments in his spare time and wander about checking them in the small hours. He also ruminated on the causes of war, writing a 10-page manuscript in 1915 titled 'Conditions of a Lasting Peace in Europe'.

After the war, Richardson worked in the meteorological office at Benson, Oxfordshire, where his work on weather prediction resulted in the book *Weather Prediction by Numerical Processes*. Testing the theory required measurements of winds at high altitudes, which at the time was not possible, however, and the calculations took longer to complete than the weather did to arrive. It was not until thirty years later, with the development of radiosonde balloons and high-speed computers, that the book was recognized as a pioneering and visionary work.

In 1920 Richardson moved to Westminster Training College and carried on with research into the upper air. It was here that a crisis of conscience arose, when later that decade he came to the attention of Sir Nelson Johnson and Sir Graham Sutton, who had an interest in the diffusion of poisonous gases. Richardson stopped his research and with much heartbreak destroyed all the work that had not been published. 'What this cost him none will ever know!', wrote his wife after his death.[26] Sutton later wrote that he was sure no arguments would have persuaded Richardson to take part in work connected with warfare, in that 'this gentle man was unshakable in his decision to conduct his life in accordance with the principles of his creed'.[27]

Richardson also turned his attention to psychology in the 1920s, obtaining a BSc in 1929 at age 48. He attempted to apply his knowledge of mathematics and physics to pure psychology, for example, estimating the intensity of mental imagery when trying to think of a particular object. This shift seems to have been at the back of Richardson's mind for decades. He recalled that while at Cambridge a friend had told him that Hermann von Helmholtz, whose pioneering research encompassed a theory of vision and theories of energy, had been a medical doctor before becoming a physicist. Richardson wrote that he felt that Helmholtz had eaten the meal of life in the wrong order, and that he would like to spend the first half of his life under the strict discipline of physics, and afterwards to apply that training to research on living things. But it was also inspired, he said, by the Quaker emphasis on social service.

Richardson did not direct his attention to problems in war and peace until 1934, feeling that the period before that was relatively tranquil. But on 18 May 1935 he published a letter in *Nature* suggesting that the rate of increase of 'preparedness-for-war' of a country is proportional to the actual 'preparedness-for-war' of its opponent. Ursula Franklin (whom we will meet later) explained that he realized that conflicting nations were similar to the hot and cold weather masses that impinge on each other causing thunderstorms and turbulence. So, he reasoned, it should be possible to see wars coming in the same way as bad weather. His work was startlingly unconventional and not without its critics: the mathematician Henry Piaggio, for example, complained that his theory didn't include the effect of intelligent aggression planned by a leader like moves in a game of chess. Nevertheless, Richardson persevered, and after retiring early to pursue his research switched from his previous approach (which was to derive his arms race model from theoretical considerations) to one that attempted to work out whether a statistical analysis of past wars would indicate any relationships that might be of practical significance. He obtained dates, numbers of casualties, and other data from the University library in Glasgow, and found that war sizes follow a power-law distribution, such that bigger wars are less common than smaller ones. Because of its unorthodoxy, the fruit of his research, *The Statistics of Deadly Quarrels,* was not published until 1960, seven years after his death.

When it comes to Richardson's religious beliefs, the picture is somewhat murky. While in France, he wrote that the conflict between science and religion that seemed so real and poignant 10 years ago no longer bothered him. He had then vigorously disbelieved in lots of things, he said, like the first chapter of Genesis, many of the miracles, the virgin birth and the resurrection. Now he felt that disbelief did not matter so much. Whether or not these things were true, religion was founded on an inner sense that we all possess to one degree or another, a sense that enables us to pray and attracts us to the lives of the saints. Echoing the view of many of our Quaker scientists, he was comfortable with doubt, happily admitting that, 'To me doubt is mostly pleasant … for fifty years I have remained comfortably in doubt as to the existence of a future life, or as to which political party is the best.'[28]

When it comes to the relationship he saw between science and his Quaker ethical convictions, however, Richardson was crystal clear, especially regarding the Quaker condemnation of war. Even from the short account of Richardson's life given here we can see that this conviction affected his actions in a variety of ways. On the one hand it meant that some areas of science were off limits, as seen in his personally costly decision to destroy research that could be used in war. On the other, it meant investing considerable time and energy in research that he thought would shed light on the causes of war.

Many of these decisions involved sacrifices. Richardson's research on the causes of war, for example, involved facing scepticism from former colleagues, and his decision to retire early to pursue it meant that the family had to live frugally. Perhaps, though, this latter sacrifice was made easier by the traditional Quaker emphasis on living simply. The future climatologist Hubert Lamb was a friend of one of Richardson's sons and often came for tea: he admitted that although the family was kind and interesting, their simple and frugal lifestyle put a brake on his own interest in joining the Society (although he did so later, after meeting other Quakers with a less austere attitude). The decision to serve in France also had repercussions, as Richardson appears to have been suffering from shellshock on his return, startling easily and often crying out.

Richardson was thus clearly a gifted scientist, something of a loner and maverick, whose research interests were shaped by his Quaker convictions about war.

Further reading

Most of the information in this section is from Oliver Ashford, *Prophet or Professor? The Life and Work of Lewis Fry Richardson* (Adam Hilger, 1985), which is available through archive.org.

2.7

Victor Paschkis: The Society for Social Responsibility in Science

Victor Paschkis (1898–1991) was born in Vienna, and earned three engineering degrees there from the Institute of Technology. Both he and his wife had Jewish ancestry, so in 1938, in the face of increasing persecution from the Nazi regime, the family emigrated to the United States. Paschkis became the technical director of the Heat and Mass Flow Analyzer Laboratory at Columbia University's School of Engineering, staying there until his retirement in 1966. His career as a mechanical engineer was long and influential. His writings on heat transfer were known around the world, and he was a key player in the development of the direct analogue computer, which, in the days before digital computers, used vast networks of resistors and capacitors to simulate complex problems, in particular in heat flow. Some of the equipment he built while at Columbia was used in a NASA moon-landing mission. Paschkis was raised as a Roman Catholic, but converted to Quakerism as an adult. During his time at Columbia he was involved with a cooperative Society of Friends community in Hidden Springs, New Jersey, and after retirement with the Society of Friends Fellowship House in Philadelphia and its Fellowship Farm in Pottstown. It is difficult to get an idea of what he was like as a person, as hardly anything has been written about him. The brief reminiscences from family members in his obituary in the *Chicago Tribune* say that he took all of life seriously and impressed on his children that it was not enough to have a pleasant life—it was important to make a difference.

Paschkis served in the First World War, but in later years was heavily influenced by the peace activist and pacifist A. J. Muste, who argued that all war was evil and there was no such thing as a just war. In 1947, Paschkis wrote an article in the Quaker journal *Friends Intelligencer* echoing Muste's views. He argued that all individuals, regardless of occupation, were personally responsible for the uses made of their work and should choose livelihoods in line with their moral values. Muste read the article, phoned Paschkis, and suggested that they should get together a group of conscientious scientists and engineers for further discussion. Paschkis duly contacted 114 scientists in the Philadelphia-New York area whom he thought might be like-minded, and in June 1948, 35 of them met at the Quaker Haverford College. Many were members of historic peace churches or worked at colleges traditionally associated with these churches, and they decided to establish an organization that would function as a support group and referral service for those who refused to participate in military projects.

While it was clear that members should refuse to work on the development of missiles and research into chemical warfare, the decisions around other activities were less obvious, as it could not always be predicted whether research might be used for military purposes in the future. It was decided that it would be up to each individual to decide, on the basis of conscience. This was in line with the Quaker emphasis on the obligation of each individual to seek God's leading through the Inner Light, although it perhaps marginalized the associated Quaker emphasis on the importance of the role of the wider Quaker community in discerning the rightness of a particular course of action. The group became known as the Society for Social Responsibility in Science (SSRS), and held its first meeting in September 1949 at the Quaker Swarthmore College, with Paschkis elected as the first president.

The SSRS made it clear that those who joined them in refusing to participate in morally objectionable practices might find their actions personally and professionally costly, with one early pamphlet advising members to maintain a moderate standard of living and accumulate a financial cushion in case they had to leave their job. Einstein was one notable scientist who clearly felt that this was a price worth paying. He himself joined and wrote a letter to the editor of *Science* urging others to do the same, arguing that it would be easier for individuals to clarify their mind in a group, and that mutual help was essential for those who faced difficulties because they followed their conscience.

Outside forces shaped the direction that the SSRS would take over the coming years. During the 1950s the emphasis was on individual action on the basis of conscience. This was a position that managed to avoid red-baiting because members refused to take political positions. The Society reached its heyday between 1959 and 1965, and through its newsletter (which had a peak readership of several thousand in 1959), letters to journal editors and conferences, provoked discussion about whether scientists themselves or the 'scientific community' was responsible for the uses made of scientific knowledge. As more overtly political forms of scientific activism opened up at the end of the McCarthy era, however, there were internal tensions. Vigils and marches, for example, against atomic testing and Vietnam, became more common, and by 1969 the SSRS's original intention of providing support for those wanting to pursue a morally acceptable career had become marginal in both the scientific and anti-war communities. It ceased operation in 1976.

Paschkis' experience in the SSRS was put to good use elsewhere though. In 1972 he was a principal founder of the Technology and Society Division of the American Society of Mechanical Engineers, and helped to form the Committee for Social Responsibility in Engineering, which later evolved into the Institute of Electrical and Electronics Engineers Society on the Social Implications of Technology, a society that is still in operation today. In 1986 he received the Award for Scientific

Freedom and Responsibility from the American Association for the Advancement of Science.

At the heart of Paschkis' crusade for social responsibility was the question of who is responsible for the use to which science is put: should it be scientists themselves or society? The physicist G. E. Owen, for example, argued that the scientist bore only the same responsibility as other members of society for the eventual use of scientific discoveries. This is partly because it is not possible to predict these uses in advance. It is also because progress in science is inevitable. If the discoveries of X-rays and radioactivity that preceded the development of the atomic bomb hadn't been made by the scientists who did make them, he argued, they would have been made by someone else instead. In contrast, the decision to make and use the atomic bomb was deliberate, military, and political.

Paschkis had counterarguments in relation to both pure science, which aims at a more complete understanding of the universe for its own sake, and applied science, in which the application is known. In relation to pure science, scientists are responsible for their work if they have a reasonable idea of what the results might be used for. In the past, pure science often did not have a sponsor but now it did, he pointed out, and it might be possible to discern what the pure science could be used for from knowing the sponsor and whether the publication of results was to be free or restricted. The Manhattan Project, for example, did not pursue pure science as an end in itself but in order to develop the atomic bomb. If it is an essential part of scientific work to start by acquainting oneself with all pertinent facts, says Paschkis, it seems entirely arbitrary to exclude social implications from the realm of pertinency. In relation to applied science, scientists should not delegate responsibility to society, because to do so would deprive them of freedom. In fact, it would make them subject to the dictates of society in much the same way as happens in an autocratic society.

Paschkis certainly practised what he preached. He turned down the opportunity to work on the Manhattan Project, and while at Columbia refused to build an analogue computer until he was assured that it would not be used for military purposes. Because he emphasized the role of individual conscience though, he did not try to impose his views on members of his lab. He recognized, for example, that while some engineers might feel that anti-missile defence systems could avert war, others might view them as threats to peace. He acknowledged that acting in accordance with one's conscience might be difficult, but in the face of the threats of society, he said, 'the pacifist scientist will feel secure in his conscience and with Luther say: "Here do I stand; I cannot other; so help me God"'.[29]

Further reading

Information on the SSRS can be found in Kelly Moore, *Disrupting Science: Social Movements, American Scientists and the Politics of the Military, 1945–1975* (Princeton University Press, 2008).

Owen and Paschkis set out their views in G. E. Owen & V. Paschkis, 'The responsibility of scientists [with reply]', *Antioch Review* 16 (1956): 161–176.

2.8

Dame Kathleen Lonsdale: Removing the causes of war

Kathleen Lonsdale (1903–1971), née Yardley, was born in County Kildare, Ireland, but her family moved to Essex, England in 1908. In the absence of classes in maths, physics and chemistry at the Ilford County High School for Girls, she studied these subjects at the County High School for Boys, winning awards in six disciplines. She went on to study physics at Bedford College for women, and her excellent performance in her BSc caught the attention of the examiner W. H. Bragg, a pioneer in X-ray crystallography. Lonsdale subsequently joined his team at University College London: 'My work was fun', she said. 'I often ran the last few yards to the laboratory.'[30] In 1928 she solved the structure of hexamethylbenzene, becoming the first person to work out the structure of an aromatic compound. Lonsdale married in 1927, and her husband bucked social norms and encouraged her to keep working. She had children in 1929, 1931 and 1934, and again kept working, this time because Bragg found her a stipend for a maid. In 1945, she and Marjory Stephenson were the first women to be elected as Fellows of the Royal Society.

Regarding her beliefs, Lonsdale was brought up as a Baptist and made a confession of faith at primary school. She had nagging doubts though, recalling that she could not really accept that a loving heavenly Father would condemn hundreds of thousands of people who couldn't believe certain things. She and her husband attended various religious services after their marriage and eventually settled on Quakerism as best reflecting their pacifist and egalitarian ideals. For her, Christianity was a way of life. It was not the unquestioning acceptance of particular beliefs such as the literal truth of the virgin birth or resurrection but rather being the kind of person that Jesus wanted his followers to be and doing the kinds of things he told them to do. Belief was still important, but it was belief in the power of goodness, justice, mercy and love, not in a wishy-washy way but to the extent that one was prepared to put them to the acid test of experiment. She said that to help students find the reality of God's presence in the face of materialism she found her faith pruned to the bone, which made it stronger. She also reported finding inspiration in the writings of Silvanus Thompson.

In 1943 Lonsdale refused to register for civil war duties. She knew she would have been exempt because of her young family, and she knew her action posed a risk to her career, but she did it anyway. The result was a sentence of a month in London's Holloway prison. Lonsdale arrived with two books—Peake's *Commentary on the Bible* and Clark's *Applied X-rays*—and the episode provides a striking vignette of an unlikely location in which one woman managed to combine science, religious reflection, and a stand for pacifism. While in Holloway, Lonsdale continued her mathematical analyses of anomalous reflections in X-ray patterns, which she later sent to the Royal Institution. She also had long conversations with prisoners and guards, and after her release became a prominent spokesperson for prison reform. Her husband told her that prison had done her the world of good. 'Husbands do say things like that', she wrote, 'but I knew what he meant, and he was right. It had made me more human, more interested in other people.'[31]

Lonsdale joined the SSRS, and her reflections on war are set out in her 1953 Swarthmore lecture *Removing the Causes of War*. Quakers have maintained throughout their history that the teaching and spirit of Jesus lead to a rejection of war in all circumstances, she says, and they dissent from the teaching of Augustine and the official church that there is such a thing as a 'just war'. She sees the causes of war as including political conceptions, philosophical tendencies, economic needs, the desire for power, racial antagonism, fear, defence and suspicion, so that any attempt to remove the causes of war must function in many directions simultaneously. Hatred also plays a role, in that all war, even the Cold War, is an expression of hatred and cruelty, often stoked by a pre-war campaign of vilification. Oppression should be met not with force but with a non-violent campaign that does not involve hatred and aims to change the heart of the oppressor: 'the one essential axiom involved is that in human nature there is something inherently good, that there is, as Friends put it, "that of God in every man", to which an appeal can be made.'[32]

Lonsdale appreciated that science played a complex role in war. On the one hand, the technological and agricultural knowledge of more developed countries could, if shared, help to remove the inequality that is one of the causes of war. On the other, war has gathered momentum only as advances in technology have provided new and more devastating weapons. Does this mean we could prevent war by halting the advances of science, she asks? Unfortunately not, as we can already destroy ourselves with the weapons we have. That said, if all scientists refused to take part in war and preparations for war, politicians would be forced to find other means of settling international disputes. While recognizing that many scientists believed that it was their duty to help defend their country, or accepted the idea of a 'just war', she insists that scientists who suffer from the illusion that armaments are the only way to prevent war should heed the words of one of their own, the eminent physicist James Prescott Joule, who warned that 'by applying itself to an improper object science may eventually fall by its own hand'.[33]

Related to Lonsdale's pacifism was her stance on scientific internationalism. Science is international by its very nature, she argued, because truth knows no frontiers and scientific knowledge has been built on the work of scientists of many nationalities. Her stance found practical expression in her attempts to increase cooperation between British and Russian crystallographers during the Cold War in the face of scepticism from colleagues. It was common at the time for letters to Russian scientists to go unanswered, and for invitations to Russian scientists to attend conferences to be ignored or declined. Lonsdale achieved a breakthrough of sorts when she joined a trip to Russia in July 1951 as one of a group of seven Quakers of varied interests. The Quakers had asked for the invitation because they wanted to meet Russian politicians, church leaders and educators to see if they could do anything to break down the barriers of misunderstanding and suspicion, and they had a wide-ranging programme that included engagements alone and as a group. Thus while the chocolatier Paul Cadbury was treated to a tour of the Red October chocolate sweets factory, Lonsdale visited the Institute of Crystallography in Moscow and the Academy of Sciences. Through impromptu lectures, a reception, and informal talks, Lonsdale managed to get to the bottom of a few of the problems. Cultural differences, for example, meant that the Russians did not deem it discourteous to ignore final dates for replies. Some mail inevitably went astray because of problems with addresses—there were, for instance, numerous 'senior librarians'. And the failure of British manufacturers to deliver equipment to Russian labs was partly due to the fact that the manufacturers had been unable to ascertain whether it would be operating in a hot or cold climate because enquiries were treated with suspicion.

In addition to relating science and religion with respect to war, Lonsdale saw a number of other points of contact. At a very concrete level, for example, science helps us to carry out the commands of Jesus to feed the hungry, clothe the naked and heal the sick on a larger scale. The development of selected seeds, fertilizers and agricultural machinery, she says, are part of practical religion just as much as they are of applied science.

On a more abstract level, there are a number of parallels. Both science and religion involve a willingness to be open-minded and to continue to think, for example. Both also require living with paradoxes. Lonsdale, like many people, found it difficult to reconcile a loving God who created the world with the cruelty of nature, but had stopped worrying about it, she said, because she had learned as a scientist how much we don't understand, and that when a scientist encounters two apparently irreconcilable ideas, they are stepping stones to new knowledge. Both also involved experiment and the thrill of discovery: 'If we knew all the answers there would be no point in carrying out scientific research. Because we do not, it is stimulating, exciting, challenging. So too, is the Christian life, lived experimentally. If we knew all the answers it would not be nearly such fun.'[34]

Lonsdale was a small, slight, bespectacled woman, with a mass of wiry hair, but the word that springs to mind when describing her is 'indomitable'. She also clearly had her feet firmly planted on the ground. The day-to-day life of Christian service has an equivalent in the working scientist's struggle with leaks and short circuits, she observed: 'We are not concerned most of the time with major problems of the meaning of life, we are simply asking God's help in our constant battle with a quick temper or a poor digestion, or a failing memory.'[35]

Further reading

Some extracts from Lonsdale's writings as selected by James Hough are available in Kathleen Lonsdale, *The Christian Life – Lived Experimentally* (Friends Service Committee, 1976). Her Swarthmore lecture is *Removing the Causes of War* (George Allen & Unwin, 1953). An account of the Quaker trip to Russia can be found in Kathleen Lonsdale (ed.), *Quakers Visit Russia* (Friends' Peace Committee, 1952) and is available online.

2.9

Ursula Franklin: The importance of structure

Ursula Franklin (1921–2016), née Martius, was born in Germany and attended school at a time when some subjects were beginning to be censored. As a result, she decided to specialize in science, reasoning that it was impossible to censor the rules of mathematics and physics. Her mother was Jewish and her father German, and they were both sent to concentration camps when war broke out. Ursula was sent to a forced labour camp. Miraculously, all three survived, and Ursula was left not with survivor's guilt but with a deep feeling of obligation to be useful in the public domain. She obtained a PhD on energy transfer in solids at the Technical University of Berlin, and in 1949 took up a postdoctoral scholarship in the Department of Physics at the University of Toronto. (She arrived with enough English to talk about nuclear physics, she said, but not enough to know what to ask for in a grocery store.) Franklin had trained in crystallography, but after realising that her work could be used for war she became a pioneer in the field of archaeometry, the dating of ancient artefacts. She taught from 1967 to 1989 in the Department of Metallurgy and Material Science at the University of Toronto, and was the first woman at the University to be named a University Professor. She was highly respected as a feminist, pacifist and public intellectual, and has been described as a Canadian national treasure.

Franklin and her husband became Quakers in the 1950s, as they wanted to stay within the Christian spectrum but, as pacifists, found it increasingly difficult to belong to a church that blessed flags and weapons. She had a vivid analogy for the arms race, describing it as being like argumentative neighbours who bought vicious dogs as deterrents, a practice that escalated until no one could come to visit, vast amounts of money were spent on dog food, and the place became 'full of dog shit'. Her perception of pacifism was broader than war, however, and included issues related to social justice: 'The commandment "Thou shalt not kill" does not speak of weapons. People can be killed by starvation. Their spirit can be killed by hopelessness, by demeaning, by depraving. One cannot allow oneself to be involved in those means.'[36]

She was also attracted to Quakerism because of its emphasis on limitless enquiry. In relation to this, she expressed bemusement at why some people got so excited about the supposed conflict between science and religion. For her, science primarily brought the realization of how little we know and how much more there is to know, and she didn't see how the accumulated results of scientific enquiry would threaten belief. Furthermore, she was far more interested in the consequences of belief than in the details, admitting to being 'sloppy' about the latter. She felt the debate lay in what scientists did with their knowledge. Her beliefs would not allow her to make an atomic bomb, for example, and the science–religion discussion should be around what sort of science society encouraged and paid for.

What held Franklin's interests in science, social justice and Quakerism together was what she referred to as 'the red thread of structure'. Whether it was the structure in atoms she saw as a crystallographer or that in social groups, she was fascinated by how the whole was shaped not only by the parts but by how the parts were connected: 'I've always been interested in how a change in structure changes properties—makes certain things possible, makes other things unlikely—whether these are chemical reactions or the interplay of people.'[37]

Franklin was influenced by the American sociologist and philosopher of technology Lewis Mumford, and thought and wrote a lot about technology. Her interest in structure led her to ask a range of questions that relate to the structure of society: What does technology do? What does it prevent us from doing? What don't we do anymore because of it? How does it affect well-being and the pursuit of peace and social justice? She pointed out, for example, that people are involuntarily included in the effects of war or pollution, that they may be excluded from a building that has an electronic barrier, and that they may get an answering machine when what they need is a hug. Furthermore, she distinguishes between 'holistic technologies' (for example, a potter's wheel or a weaver's loom), which are used by artisans who know each step of the production process and are in control, and 'prescriptive technologies'. Prescriptive technologies are associated with the division of labour and with the demand that *things have to fit.* With these latter technologies, management and control become legitimate, in that both the part that is being made and the worker have to comply with certain rules and standards.

She also thought and wrote about feminism in both science and society. Again, her interest in structure is apparent, for example, in her appreciation that feminism embraced non-hierarchical practices based on cooperation, respect and horizontal solidarity. She argued that women should develop new ways of doing science that were less reductionist and paid attention to context, which she felt were strengths of the feminist approach to life in general. As a female scientist in what was a male-dominated field, she had first-hand experience of some of the difficulties faced by women though. She had very practical advice for women scientists, suggesting that if their male colleagues were acting like 'jerks' they should take the approach of an anthropologist studying a strange tribe: observe the tribe's customs and attitudes with detachment and consider publishing your field observations.

One of the ways Franklin exhibited some of these convictions in practice was her involvement in the early 1960s in studies of radioactive fallout. The organization Voice of Women was trying to prevent the testing of nuclear weapons in the atmosphere, but very few of the members had any scientific training. Franklin realized that strontium-90 was saturating the grass that cows were eating and thence making its way into babies. She was instrumental in organizing the collection of baby teeth to ascertain the presence of strontium, and in the process educated women, children, and the government about fallout. It was situations like this that gave rise to her frequent use of the image of a scientist as a 'citizen with a toolbox'. In many cases, scientists develop their toolbox at the expense of developing their citizenship, she explains; however, the lay public may be good citizens but have a limited toolbox. One of the joys of her professional life was working with citizen groups on issues like this, and she enthused that she had no problem in explaining quite complex chemical and physical phenomena to women, who thrived in a non-competitive atmosphere outside compulsory education and shared their new knowledge of science in a particular context at public hearings.

Finally, Franklin appreciated the silence that forms the bedrock of Quaker meetings. Silence allows for the possibility of something that is not only unforeseen but also unforeseeable to happen, she said. If one's openness to the unexpected, the unprogrammed and unprogrammable is reduced, then creativity goes, calmness goes, and a sense of proportion goes. She felt it was important to protect the soundscape and possibility of silence in the world at large as a common good, in much the same way as we might protect a landscape. We should question, for example, the 'slop' that is played in elevators, and the mood manipulation through music that goes on in sports arenas.

In her eighties, Franklin reflected that she had tried to wrestle with just one fundamental question over the previous four decades: 'How can one live and work as a pacifist in the here and now and help to structure a society in which oppression, violence and wars would diminish and co-operation, equality, and justice would rise?'[38]

Further reading

There are two anthologies of Franklin's thought: Ursula Franklin, *The Ursula Franklin Reader: Pacifism as a Map* (Between the Lines, 2006), and Ursula Franklin, *Ursula Franklin Speaks: Thoughts and Afterthoughts* (McGill-Queens University Press, 2014).

2.10

Dame Jocelyn Bell Burnell: Quakerism and science as bedfellows

Susan Jocelyn Bell was born in Northern Ireland in 1943, but had most of her secondary school education at the Quaker Mount boarding school in York. She was good at physics, and, after reading some books from the local library, decided at 15 that she would become an astronomer. More specifically, realizing that she couldn't function properly without a good night's sleep, she concluded that she should specialize in radio astronomy, which, unlike optical astronomy, doesn't require observations to be made at night. Her achievements as an astronomer are significant, but alongside research positions she has held a variety of other science-related jobs and prestigious roles, including Chair of the Physics Department at the Open University, Dean of the Science Faculty at the University of Bath, and terms as President of the Royal Astronomical Society and the Institute of Physics. She has also won numerous awards, including in 2018 the Special Breakthrough Prize in Fundamental Physics, which recognizes the world's top scientists who are asking 'big' questions about the nature of the Universe. She married Martin Burnell in 1968, but the couple later divorced.

Scientifically, Bell Burnell is best known for her discovery of pulsars while a graduate student at Cambridge University in the late 1960s. She helped to build a radio telescope, and as results started to come in found a curious signal taking up only 5 mm in 500 m of chart paper. It could easily have been overlooked or dismissed, but Bell determined that it was a string of regular pulses just over a second apart. Over the coming months she established that this signal wasn't of human origin or due to 'noise' in the telescope operating system. Rather, it turned out to be the signature of a previously unknown class of stars. Pulsars, as they were subsequently dubbed, are incredibly small and dense stars, with a mass of around one and a half times that of the Sun but the radius of a city. They send out a beam of radiation that appears to blink on and off, sometimes hundreds of times a second, because the stars are rotating and the beams are not always pointing at the Earth. They thus act a bit like cosmic lighthouses, and astronomers can make use of them to study space-time and what may lie in the regions between stars. A map of pulsars in relation to the Sun was even used on plaques on the *Pioneer* spacecrafts in the 1970s to show our location in the galaxy.

The importance of the discovery, but not Bell's role in it, was recognized by the Nobel committee, who in 1974 awarded the Nobel Prize in physics to her supervisor, Anthony Hewish, and Martin Ryle. The decision has been roundly criticized since, but Bell herself has always accepted the decision with equanimity, acknowledging that at the time it was usual for credit to be given to the leaders of scientific projects with little recognition of the role played by research students. Her reminiscences about her treatment by the press, when the discovery sparked wide interest, are less phlegmatic, however, as she recalls how uncomfortable she was when Hewish was asked about the astrophysical significance of the discovery but she was asked about her boyfriends and vital statistics.

When it comes to relating astronomy and Quakerism, Bell Burnell describes the two as long-standing and comfortable bedfellows, but that it is the Quaker bedfellow that has done the accommodating. In other words, her scientific understanding has emerged unscathed by its contact with Quakerism, but her Quakerism has bent to fit in with what she has learnt as an astronomer. At times she has seemed reluctant to share just how 'accommodating' her Quakerism had been, worrying that others might find her personal theology shocking and wanting the freedom to evolve her beliefs without setting them down for posterity. She did, however, give the Swarthmore lecture *Broken for Life* in 1989 on the theme of suffering, and the James Backhouse lecture in Australia, published as *A Quaker Astronomer Reflects,* in 2013. Both provide some insights into her beliefs, misgivings, and approach to relating Quakerism and science.

Her two main theological conclusions are first that God is not the creator of the universe, and second that God is not in control of the universe. The first owes much to her astronomically informed perception that the physical universe is purposeless: meteorites crash into planets and extinguish life; stars expand and engulf planets; galaxies collide and disrupt planetary systems. The future of the universe, as stars run out of fuel, is unremittingly 'black and bleak'. The second is a response to the perennial theological chestnut of how a loving and powerful God can allow suffering. In *Broken for Life* she approaches this problem using the scientific method, namely by suggesting a number of models and then interrogating them. She finds most explanations inadequate. Do we suffer because we have been bad? No—many children suffer. Does suffering come to ennoble us? No—it does sometimes, but this is an accidental spin-off. Perhaps then, she reasons, the problem is with the starting assumptions of a God who is both loving and powerful. Loath to abandon the idea of a loving God, Bell Burnell instead dispenses with the idea of God in control. While this means that God can no longer be blamed for suffering, it does also mean that God cannot take the credit for things that go well, or be asked to intervene.

Bell Burnell's reluctance to abandon a loving God is not just wishful thinking, but arises from her experience. She talks of having a sense of the numinous, which would earn her the label of 'mystic' in some societies, and of her experiential knowledge of God: 'a God of inspiration, of creativity; a God we can sense in the silence of a gathered Quaker meeting ... a God who supports us, cares for us, grieves with us, empowers us and acts through us in the world'.[39]

As we have seen throughout this book, though, there are many ways to relate science and religion that aren't specifically theological. For Bell Burnell this is most apparent in her treatment of her students, a treatment that picks up on the Quaker testimony of equality as well as arising from her personal experience of being a female astronomer in a male-dominated profession. In a 2008 talk at the Annual Conference of Higher Education, for example, she ponders how we can help fundamentalist students who arrive in the West to be more open-minded, and how we can find gentler ways than a public display of grades to let students know how they have performed. In particular, she has been concerned with how women can be encouraged to pursue careers in science, and how they can be supported. The Breakthrough Prize came with $3 million, which she donated to a scheme being run by the Institute of Physics to fund graduate students who are underrepresented in physics, particularly women, people of colour, refugees and people from underrepresented socioeconomic groups. The aim is to improve diversity in physics departments in the UK and Ireland.

Finally, Bell Burnell points to the similarities between science and Quakerism when it comes to evolving one's beliefs in the light of experience. Quakers expect that spiritual learning will continue throughout life, she points out, so naturally our beliefs will evolve. Similarly, scientific theories are rejected, modified, or generally accepted as new evidence emerges. Both these processes mean being comfortable with doubt, and Bell Burnell ends *A Quaker Astronomer Reflects* with one of her favourite quotes, from Rainer Maria Rilke: 'Be patient towards all that is unsolved in your heart ... do not now seek the answers which cannot be given because you would not be able to live them ... live the questions now.'[40]

Further reading

The two lectures referred to are Jocelyn Bell Burnell, *A Quaker Astronomer Reflects* (Australia Yearly Meeting of the Religious Society of Friends, 2013), and Jocelyn Bell Burnell, *Broken for Life,* Swarthmore Lecture 1989 (Quaker Home Service, 1989). There is an interview with her by Toni Felder, 'Q&A: Pulsar pioneer Jocelyn Bell Burnell', *Physics Today* 30 Jan (2019).

Chapter 3

Relating Quakerism and science

In the first part of this book, we saw how Quakers developed a strong foundation for scientific pursuits. This was partly because restrictions on their choice of profession and social activities channelled them towards careers and pastimes that involved science, and partly because they found opportunities for spiritual growth in various scientific activities. We also saw how the Manchester Conference set the tone for a Quaker revival, in which Liberal Quakers emphasized social engagement and the importance of science. They declared that the Bible should not be taken literally and enthused that science would correct doctrine.

We then met ten Quakers who wove together their faith and science in their lives in a way that bore the imprints of Manchester. They all had different personalities and pursued different careers, and therefore lived out the relationship between their faith and scientific practice in different ways. Thompson had a strong sense of social responsibility, and this was reflected in his work at Finsbury Technical College. Eddington's familiarity with the latest discoveries in physics led him into the abstract realms of philosophy. Jones was a charismatic leader, and thus it was perhaps inevitable that his use of William James' 'more' would find its way into the heart of Liberal Quakerism. Starbuck was a complex character, seeking to understand his own religious experience and that of others. Brinton's scientific disposition led him to view Quakerism as a method and uncover the Quaker testimonies. Richardson craved solitude, and his particular contribution to pacifism was a solitary investigation into the mathematics of war and peace. Paschkis, by contrast, was clearly community-minded: he lived in a Quaker community in retirement, and brought people together through the SSRS. Lonsdale, in her voluntary imprisonment, demonstrated that she was prepared to stand up for her values and beliefs and, in the revered Quaker phrase, 'speak truth to power'. Franklin, who had experienced structural evil first-hand under the Nazis, explained that what held together her interests in crystallography and religion was 'the red thread of structure', such that injustice could often be alleviated through the consideration of structures. Finally, Bell Burnell accommodated her beliefs to her religious experience and knowledge of astronomy.

In spite of these different expressions of science and faith, readers may have been struck by a number of themes that kept cropping up. This is perhaps not surprising, as there was a Quakerly web of connections among these scientists, ensuring that ideas circulated freely. Thompson and Richardson both went to Bootham, for example; Jones taught Brinton and knew Starbuck; Eddington was steeped in the ideas of Thompson and Jones; Franklin was familiar with Richardson's work; and Lonsdale appreciated Thompson's writings and joined the SSRS founded by Paschkis.

One such theme is the importance of social responsibility, which arises again and again: we saw how early Quakers used their scientific discoveries to argue for abolitionism, and how Victorian Quakers worried that the practice of vivisection would damage virtue in society. Likewise, many of the twentieth-century Quakers we considered were concerned with the social problems of their time. For example, we saw how Starbuck aspired to eradicate moral ugliness, how Paschkis founded a number of societies to encourage social responsibility among those working in science and technology, how Lonsdale argued that technology should be shared with developing nations, and how Franklin thought of scientists as citizens with a toolbox. Related to this issue of social responsibility is that many of these Quakers were wrestling with ethical questions posed by war. Thus we saw how Richardson destroyed unpublished work that he thought might be used to investigate the diffusion of poisonous gases, how Eddington stood against the prevailing wartime attitude of distrust to work with Einstein, how Lonsdale went to prison for refusing to register for war duties, and how Franklin sought evidence that radioactive material from weapon testing was entering the food chain.

A second theme that stands out is the conviction that it is important to be open to new expressions of faith. Thus Thompson enthused that science would clear away incorrect doctrine; Jones reformulated the concept of the Inner Light using ideas from William James; Brinton sought to link the Inner Light with prevalent ideas about the evolution of the universe; and Bell Burnell moved away from ideas of God as an all-powerful creator. This willingness to keep on searching for truth is also apparent in the way that many of these scientists expressed their comfort with doubt. We saw that Thompson felt that the craving for certitude was not a sign of spiritual health; Richardson said doubt was 'mostly pleasant'; Lonsdale took encouragement from the fact that irreconcilable ideas in science were stepping stones to new knowledge; and Bell Burnell ended her reflections with a quotation about living with unanswered questions.

A third theme that emerges is that these twentieth-century scientists stressed the importance of experience. Thus Thompson formulated spiritual laws based on experience; Jones wrote prolifically about religious experience through the lens of mysticism; Starbuck structured his questionnaires around religious experience; Richardson said that what mattered was the inner sense on which religion was founded; and Bell Burnell kept hold of the idea of God as benevolent based on her experience.

Finally, many of these scientists saw no sharp distinction

between religious and scientific approaches to life. This is expressed in the way that they saw parallels between the methods used in their scientific practice and Quakerism. We saw that as early as the Manchester Conference, Thompson had pointed to the fact that the habit of accurate thought and speech in which 'yea means yea and no more' was fundamental to both science and Quakerism, and we discussed how Brinton saw parallels between Quakerism and the scientific method. We also looked at how Eddington saw the same goal of continual seeking in Quakerism and science, and how Starbuck was convinced that psychology could be used in the service of religion.

The next question we might want to ask, then, is whether anything lies behind these themes. I think that the idea of the Inner Light, which is so central to Quakerism, is helpful here. Although Rufus Jones radically reformulated the way Liberal Quakers thought about the Inner Light, he wrote that it has generally been understood in three main ways: first, as a divine life resident in the soul; second, as a source of guidance and illumination; and third, as the ground of spiritual experience.

So, first, we can say that the idea that the Inner Light is a divine life in the soul, such that there is 'that of God' in everyone, has moral implications. As Franklin puts it, it 'provides a code of conduct for every human being, because if we are equally important to God, then who are we to classify people into those who matter and those who don't?'[1] This conviction goes some way to explaining the emphasis on social responsibility.

Second, the Inner Light as a source of illumination conjures up images of seeing and understanding things clearly, of making new discoveries, and of dispelling ignorance, so this ties in with the theme of being open to new expressions of faith. These are all characteristics we might also associate with science of course. We should therefore perhaps not be surprised that people who are naturally curious and want to discover things for themselves would be attracted to both Quakerism and a career in science. Lonsdale and Eddington both talked about the joy and thrill of seeking new knowledge, for example, and Franklin said that one of the things that attracted her to Quakerism was its emphasis on limitless enquiry.

And third, the idea that the Inner Light is the ground of spiritual experience is, Jones enthuses, in line with the modern realization that the criterion of truth is found in the nature of consciousness, not somewhere else. A person knows they have received forgiveness because it is witnessed within, he says, not because 'some man in sacred garb has announced it, or because I have read in a book that such an experience might be mine'.[2] This approach to religious truth is at the heart of Quakerism, so it is not surprising that experience is a common theme in the religious writings of all our scientists.

This emphasis on experience also goes some way to explaining the fourth theme, namely that there is no distinction between religious and scientific approaches to life. Just as Quakers want to experience religious teachings for themselves rather than at second-hand, scientists conduct their own experiments and perform their own calculations to test theories (at least initially, until theories have been tested so often that they are generally accepted). In his book *Truth of the Heart*, the Quaker writer Rex Ambler sees this parallel as going back to the beginning of Quakerism. George Fox was doing for religion what Galileo and Newton were doing for science, he says. Fox was rejecting knowledge that was passed on by authorities and testing the matter for himself through experience.

It is also interesting to speculate on what these Quaker scientists would have made of the problems facing us today. I suspect that Lonsdale would have argued vigorously for climate justice and the sharing of green technologies, and that Franklin would have expressed reservations about cashier-less supermarkets and about how our addiction to mobile phones can destroy inner silence. I also suspect that Starbuck would have been uncomfortable about the use of psychology in marketing. Some issues are more ambiguous though. Would Paschkis have refused to work on equipment for fracking, I wonder?

Richard Tuckett, whom we have not discussed, is one interesting example of a present-day scientist trying to apply Quaker principles to the problems caused by climate change. He is a Quaker and a chemist, and in a recent scientific article on climate change broke with academic norms by referencing the Quaker testimonies.[3] One of the ways to combat and deal with climate change, he says, is to live by the traditional testimonies of social justice and equality, and the more recent one of sustainability. The importance of living sustainably is obvious in relation to not over-using resources and cutting our carbon emissions through driving and flying less, for example. The testimonies of social justice and equality have led Quakers (and others by different means of course) to the conclusion that because it is industrialized countries that have landed the world in its current climate crisis, it is these countries that should take the lead in resolving it. While acknowledging that religion and morality are outside the remit of scientific papers, Tuckett suggests that the problem of climate change is so different from other scientific issues that this 'rule' should be put aside. Interestingly, and complementary to this view, TORCH (The Oxford Research Centre in the Humanities) recently posted a blog on academic activism. The author argued that given there is so little time to act on climate change, academics should reconsider the traditional view that it is not possible to be committed to a particular agenda while maintaining scholarly objectivity. In fact, they should use their expertise to respond to escalating climate threats.[4]

This book has covered a lot of ground and some complex ideas, but I hope that it has shown that the relationship between Quakerism and science is rich and multi-faceted. There are cultural, social and religious reasons why Quakers embraced science prior to the twentieth century, and the desire to modernize the Society that was apparent at the Manchester Conference ensured that this foundation was built upon. Because of their emphasis on deeds rather than creeds and their understanding of the Inner Light, Quakers have a distinctive contribution to make to discussions about the relationship between science and religion. Quaker scientists today thus have a heritage that includes taking a stand for the right use of science and seeing a close connection between the scientific and spiritual worlds. In the face of the current climate crisis, these insights may well prove invaluable.

Notes

Chapter 1

1 Geoffrey Cantor, *Quakers, Jews and Science* (Oxford Scholarship Online, 2005), 56.

2 George Fox, *George Fox: An Autobiography*, ed. by Rufus Jones (Ferris & Leach, 1903), Kindle loc. 847.

3 Cantor, *Quakers, Jews and Science*, 236.

4 John Woolman, *The Journal of John Woolman, Quaker* (Ignacio Hills Press, [1774]), Kindle loc. 2250.

5 Maria Hack, *Harry Beaufoy, Or, the Pupil of Nature* (Thomas Kite, 1828), 97.

6 Priscilla Wakefield, *Mental Improvement* (Darton, Harvey and Darton, 1810), 152.

7 Rufus Jones, *The Trail of Life in the Middle Years* (MacMillan, 1934), 64–65.

8 Society of Friends, *Manchester Conference* (Headley Brothers, 1895), 3–32.

9 Society of Friends, *Manchester Conference* (Headley Brothers, 1895), 203–247.

10 Jesse H. Holmes, 'To the scientifically-minded'. *Friends Journal* June 1992, 22–23 [First published in *Friend's Intelligencer*, Feb 11, 1928].

Chapter 2

1 Fox et al. 1660, quoted in Rachel Muers, *Testimony* (SCM Press, 2015), 55.

2 Silvanus P. Thompson, *The Quest for Truth* (Headley Brothers, 1915), 92.

3 Silvanus P. Thompson, *A Not Impossible Religion* (William Clowes and Sons, 1918), xi.

4 Thompson, *A Not Impossible Religion*, 187.

5 Hannah Gay and Anne Barrett. 'Should the cobbler stick to his last? Silvanus Phillips Thompson and the making of a

scientific career', *British Journal for the History of Science,* 35 (2002): 151–186, p. 168.

6 Arthur Stanley Eddington, *Science and the Unseen World,* Swarthmore Lecture 1929 (Quaker Books, 2007), 20.

7 Eddington, *Science and the Unseen World,* 24.

8 Eddington, *Science and the Unseen World,* 15.

9 Anthony Manousos, *Howard and Anna Brinton* (QuakerBridge, 2013), 47.

10 Rufus Jones, *Finding the Trail of Life* (George Allen & Unwin, 1926), 138.

11 Rufus Jones, *The Trail of Life in College* (MacMillan, 1929), 64–65.

12 William James, *The Varieties of Religious Experience* (Penguin, [1902] 1982), 512.

13 Jones, *Life in College,* 180.

14 Edwin Starbuck, 'Religion's use of me', in *Religion in Transition,* ed. V. Ferm (George Allen & Unwin, 1937), 202.

15 Starbuck, 'Religion's use of me', 215.

16 Edwin Starbuck, *Psychology of Religion* (Walter Scott, 2011), 22–23.

17 Starbuck, 'Religion's use of me', 228.

18 Starbuck, 'Religion's use of me', 209.

19 Starbuck, 'Religion's use of me', 255.

20 Anthony Manousos, *Howard and Anna Brinton* (Quakerbridge, 2013), 30.

21 Howard Brinton, *Guide to Quaker Practice,* Pendle Hill Pamphlet 20 (Pendle Hill Publications, [1942] 2006), Kindle loc. 232.

22 Manousos, *Howard and Anna Brinton,* 226.

23 Brinton, *Creative Worship,* 32.

24 Brinton, *Creative Worship,* 55.

25 Oliver Ashford, *Prophet or Professor? The Life and Work of Lewis Fry Richardson* (Adam Hilger, 1985), 43.

26 Ashford, *Prophet or Professor?,* 127.

27 Ashford, *Prophet or Professor?*, 128.
28 Ashford, *Prophet or Professor?*, 159.
29 Victor Paschkis, 'The scientist's responsibility: A pacifist view', *Bulletin of the Atomic Scientists*, Sep (1955), 266.
30 Kathleen Lonsdale, *The Christian Life – Lived Experimentally* (Friends Home Service Committee, 1976), 9.
31 Lonsdale, *Christian Life*, 12.
32 Kathleen Lonsdale, *Removing the Causes of War*, Swarthmore Lecture 1953 (George Allen & Unwin, 1953), 68.
33 Lonsdale, *Removing the Causes of War*, 17.
34 Lonsdale, *Christian Life*, 37.
35 Lonsdale, *Christian Life*, 21.
36 Ursula Franklin, *Ursula Franklin Speaks* (McGill-Queen's University Press, 2014), 58.
37 Franklin, *Ursula Franklin Speaks*, 159.
38 Ursula Franklin, *The Ursula Franklin Reader: Pacifism as a Map* (Between the Lines, 2006), 39.
39 Jocelyn Bell Burnell, *A Quaker Astronomer Reflects* (Australia Yearly Meeting of the Religious Society of Friends, 2013), Kindle loc. 661.
40 Bell Burnell, *A Quaker Astronomer*, Kindle loc. 716.

Chapter 3

1 Ursula Franklin, *Ursula Franklin Speaks* (McGill-Queens University Press, 2014), 30.
2 Rufus Jones, *Social Law in the Spiritual World* (Headley Brothers, 1904), 172.
3 Richard Tuckett, 'Climate Change and Global Warming: What can we do, what should we do?', in *Reference Module in Earth Systems and Environmental Sciences* (Elsevier, 2018).
4 Amanda Power, 'The future of academic activism?' Retrieved from: https://www.torch.ox.ac.uk/article/the-future-of-academic-activism.

Bibliography

Ambler, Rex. *Truth of the Heart – An Anthology of George Fox.* Quaker Books, 2001.

Angell, Stephen W. & Dandelion, Pink (eds.). *The Oxford Handbook of Quaker Studies.* OUP, 2015.

Ashford, Oliver. *Prophet or Professor? The Life and Work of Lewis Fry Richardson.* Adam Hilger, 1985.

Bell Burnell, Jocelyn. *Broken for Life,* Swarthmore Lecture 1989. Quaker Home Service, 1989.

Bell Burnell, Jocelyn. *Heavens Above and Heaven on Earth.* Lecture given to the Friends Association of Higher Education, June 2008. Retrieved from https://www.woodbrooke.org.uk/resource-library/fahe-2008-jocelyn-burnell/ [Accessed 11 November 2021].

Bell Burnell, Jocelyn. *A Quaker Astronomer Reflects.* Australia Yearly Meeting of the Religious Society of Friends, 2013.

Brinton, Howard. *Creative Worship,* Swarthmore Lecture 1931. George Allen & Unwin, 1931.

Brinton, Howard. *Guide to Quaker Practice,* Pendle Hill Pamphlet 20. Pendle Hill Publications, [1942] 2006.

Brinton, Howard. *Evolution and the Inward Light: Where Science and Religion Meet,* Pendle Hill Pamphlet 173. Pendle Hill Publications, 1970.

Brinton, Howard. *Friends for 300 Years.* Pendle Hill Publications, [1952] 1994.

Cantor, Geoffrey. *Quakers, Jews and Science.* Oxford Scholarship Online, 2005.

Cantor, Geoffrey. 'Quakers and Science'. In *The Oxford Handbook of Quaker Studies,* pp. 520–534, ed. by Stephen W. Angell and Pink Dandelion. OUP, 2015.

Dandelion, P. *The Quakers: A Very Short Introduction.* OUP, 2008.

Dudiak, J. and Rediehs, L. 'Quakers, Philosophy and Truth'. In

The Oxford Handbook of Quaker Studies, pp. 507–519, ed. by Stephen W. Angell and Pink Dandelion. OUP, 2015.

Eddington, Arthur Stanley. *Science and the Unseen World*, Swarthmore Lecture 1929. Quaker Books, 2007.

Eddington, Arthur Stanley. *The Nature of the Physical World*. MacMillan, 1928.

Eddington, Arthur Stanley. 'Physics and philosophy'. *Philosophy* 8 (1933): 30–43.

Felder, Toni. 'Q&A: Pulsar pioneer Jocelyn Bell Burnell'. *Physics Today*, 30 Jan 2019.

Fox, George. *George Fox: An Autobiography*, ed. by Rufus Jones. Ferris & Leach, 1903.

Franklin, Ursula. *The Ursula Franklin Reader: Pacifism as a Map*. Between the Lines, 2006.

Franklin, Ursula. *Ursula Franklin Speaks: Thoughts and Afterthoughts*. McGill-Queens University Press, 2014.

Gay, Hannah & Barrett, Anne. 'Should the cobbler stick to his last? Silvanus Phillips Thompson and the making of a scientific career'. *British Journal for the History of Science* 35 (2002): 151–186.

Glaholt, Hayley Rose. 'Vivisection as war: the "moral diseases" of animal experimentation and slavery in British Victorian Quaker ethics'. *Society and Animal* 20 (2012): 154–172.

Grant, Rhiannon. *Quakers Do What! Why?* Christian Alternative, 2020.

Grieg, James. *Silvanus P. Thompson: Teacher*. Her Majesty's Stationery Office, 1979.

Hack, Maria. *Harry Beaufoy, Or, the Pupil of Nature*. Thomas Kite, 1828.

Hodgkin, Dorothy. 'Kathleen Lonsdale, 28 January 1903 – 1 April 1971'. Biographical Memoirs of Fellows of the Royal Society 21 (1975): 447–484.

Holmes, Jesse. 'To the scientifically-minded', *Friends Journal* June (1992): 22–23 [First published in *Friend's Intelligencer*,

Feb 11, 1928].

Holt, Helen. *Mysticism and the Inner Light in the Thought of Rufus Jones, Quaker*. Brill, 2022.

James, William. *The Varieties of Religious Experience*. Penguin, [1902] 1982.

Jones, Rufus. *Social Law in the Spiritual World*. Headley Brothers, 1904.

Jones, Rufus. *Finding the Trail of Life*. George Allen & Unwin, 1926.

Jones, Rufus. *The Trail of Life in College*. MacMillan, 1929.

Jones, Rufus. *The Trail of Life in the Middle Years*. MacMillan, 1934.

Kavanagh, Jennifer. *Practical Mystics*. Christian Alternative, 2019.

Leach, Camilla. 'Religion and rationality: Quaker women and science education 1790–1850'. *History of Education* 35 (2006): 69–90.

Lonsdale, Kathleen. *The Christian Life – Lived Experimentally*. Friends Service Committee, 1976.

Lonsdale, Kathleen. *Removing the Causes of War*. George Allen & Unwin, 1953.

Lonsdale, Kathleen (ed.). *Quakers Visit Russia*. Friends' Peace Committee, 1952.

Manousos, Anthony. *Howard and Anna Brinton: Re-inventors of Quakerism in the Twentieth Century*. Quakerbridge, 2013.

McGrath, Alister E. *Science and Religion: A New Introduction*. Wiley Blackwell, 2010.

Moore, Kelly. *Disrupting Science: Social Movements, American Scientists and the Politics of the Military, 1945–1975*. Princeton University Press, 2008.

Muers, Rachel. *Testimony*. SCM Press, 2015.

Owen, G.E. & Paschkis, Victor. 'The responsibility of scientists [with reply]', *Antioch Review* 16 (1956): 161–176.

Paschkis, Victor. 'The scientist's responsibility: A pacifist view',

Bulletin of the Atomic Scientists Sep (1955): 265–266.

Power, Amanda. 'The future of academic activism?' Retrieved from: https://www.torch.ox.ac.uk/article/the-future-of-academic-activism.

Society of Friends, *Report of the Proceedings of the Conference in Manchester*. Headley Brothers, 1895.

Stanley, Matthew. *Practical Mystic*. University of Chicago Press, 2007.

Starbuck, Edwin. 'Religion's use of me'. In *Religion in Transition*, ed. by V. Ferm, pp. 201–260. George Allen & Unwin, 1937.

Starbuck, Edwin. *The Psychology of Religion*. Walter Scott, 1911.

Thompson, Silvanus P. *The Quest for Truth*. Headley Brothers, 1915.

Thompson, Silvanus P. *A Not Impossible Religion*. William Clowes and Sons, 1918.

Tuckett, Richard. 'Climate change and global warming: what can we do, what should we do?' In *Reference Module in Earth Systems Environmental Sciences*. Elsevier, 2018. Retrieved from: https://doi.org/10.1016/B978-0-12-409548-9.11355-7 [Accessed 18 August 2021].

Valentine, Lonnie. 'Quakers, war, and peacemaking'. In *The Oxford Handbook of Quaker Studies*, ed. by Stephen W. Angell and Pink Dandelion, pp. 363–376. OUP, 2015.

Vining, Elizabeth. *Friend of Life*. Forgotten Books, 1958.

Wakefield, Priscilla. *Mental Improvement*. Darton, Harvey and Darton, 1810.

Woolman, John. *The Journal of John Woolman, Quaker*. Ignacio Hills Press, [1774].

Websites

www.quaker.org.uk

www.quakersintheworld.org/quakers-in-action/what/science

https://womenyoushouldknow.net/astronomy-maria-mitchell/

Glossary

Inward/Inner Light

The Light has always been a central concept in Quakerism, but it has generally resisted precise definition, and understandings of it have changed over time. Broadly speaking, it is a way of expressing the idea that there is 'that of God' within a person. In the seventeenth century, Quakers associated the Light with Christ and defended their position biblically (for example, quoting passages in John's prologue about Christ being the light of the world). In the eighteenth century, Quietist Quakers emphasized the inner witness of the Light at the expense of the Bible and reason. In the nineteenth century, those Quakers influenced by evangelicalism viewed Scripture as more authoritative than the Light and associated the Light with the Holy Spirit, but others saw this trend as threatening the very essence of Quakerism. At the beginning of the twentieth century, Rufus Jones radically reshaped the way in which Liberal Quakers thought about the concept, claiming that psychology had shown that the Light was not separate from humans, like a candle in a lantern, but was an integral part of human nature. He used the terminology 'Inner Light', rather than the 'Inward Light' favoured by earlier Quakers.

Liberal Quakerism

Liberal Christianity arose in the nineteenth century and interprets Christianity taking into account science, biblical criticism, ethics, and personal religious experience. Related to this, Liberal Quakerism became the dominant form of Quakerism in Britain following the Manchester Conference of 1895, where Liberal speakers recognized that Christianity should adapt to fit contemporary thought, acknowledged that it was impossible to reconcile a literal reading of Genesis with advances in scientific knowledge, and stressed that theology had to be based on an appeal to an experience of God. Quakerism was in fact well placed to adapt to Liberal Christianity. Although the motivation of seventeenth-century Quakers was very different from that of nineteenth-century Liberal Christians, the traditional Quaker emphasis on the importance of experience and continual revelation through the Light fitted the new paradigm well. William James, in fact, remarked in *The Varieties of Religious Experience* that Liberal Christians were reverting to what George Fox and the Quakers had long ago assumed. Present-day Liberal Quakerism is accepting of a variety of belief systems, such that many Liberal Quakers identify, for example, as Muslim Quakers, or Buddhist Quakers, or non-theist Quakers.

Evangelical Quakerism

Evangelical Christianity emphasizes the authority of the Bible and the importance of Christ's death on the cross, which makes a relationship with God possible. It spread through Britain and North America in the late eighteenth and early nineteenth centuries partly through revival meetings, at which individuals were encouraged to make a declaration of faith that enabled them to be 'born again'. While Quakers, as Christians, were not exempt from the influence of evangelicalism, their traditional emphasis on the Inward Light gave rise to some tensions. In America, Quakerism split in 1827 in what became known as the 'Great Separation': so-called Orthodox Quakers emphasized the primacy of Scripture, whereas Hicksites (after Elias Hicks, who led opposition to more evangelical formulations of faith) emphasized the primacy of the Inward Light. Today, over 80 per cent of Quakers worldwide are evangelical, with this strand of Quakerism being particularly strong in America and Kenya, but not in Britain.

THE NEW OPEN SPACES

Throughout the two thousand years of Christian tradition there have been, and still are, groups and individuals that exist in the margins and upon the edge of faith. But in Christianity's contrapuntal history it has often been these outcasts and pioneers that have forged contemporary orthodoxy out of former radicalism as belief evolves to engage with and encompass the ever-changing social and scientific realities. Real faith lies not in the comfortable certainties of the Orthodox, but somewhere in a half-glimpsed hinterland on the dirt track to Emmaus, where the Death of God meets the Resurrection, where the supernatural Christ meets the historical Jesus, and where the revolution liberates both the oppressed and the oppressors.

Welcome to Christian Alternative... a space at the edge where the light shines through.

If you have enjoyed this book, why not tell other readers by posting a review on your preferred book site.

Recent bestsellers from Christian Alternative are:

Bread Not Stones

The Autobiography of An Eventful Life

Una Kroll

The spiritual autobiography of a truly remarkable woman and a history of the struggle for ordination in the Church of England.

Paperback: 978-1-78279-804-0 ebook: 978-1-78279-805-7

The Quaker Way

A Rediscovery

Rex Ambler

Although fairly well known, Quakerism is not well understood. The purpose of this book is to explain how Quakerism works as a spiritual practice.

Paperback: 978-1-78099-657-8 ebook: 978-1-78099-658-5

Blue Sky God

The Evolution of Science and Christianity

Don MacGregor

Quantum consciousness, morphic fields and blue-sky thinking about God and Jesus the Christ.

Paperback: 978-1-84694-937-1 ebook: 978-1-84694-938-8

Celtic Wheel of the Year

Tess Ward

An original and inspiring selection of prayers combining Christian and Celtic Pagan traditions, and interweaving their calendars into a single pattern of prayer for every morning and night of the year.

Paperback: 978-1-90504-795-6

Christian Atheist

Belonging without Believing

Brian Mountford

Christian Atheists don't believe in God but miss him: especially the transcendent beauty of his music, language, ethics, and community.

Paperback: 978-1-84694-439-0 ebook: 978-1-84694-929-6

Compassion Or Apocalypse?

A Comprehensible Guide to the Thoughts of René Girard

James Warren

How René Girard changes the way we think about God and the Bible, and its relevance for our apocalypse-threatened world.

Paperback: 978-1-78279-073-0 ebook: 978-1-78279-072-3

Diary Of A Gay Priest

The Tightrope Walker

Rev. Dr. Malcolm Johnson

Full of anecdotes and amusing stories, but the Church is still a dangerous place for a gay priest.

Paperback: 978-1-78279-002-0 ebook: 978-1-78099-999-9

Do You Need God?

Exploring Different Paths to Spirituality Even For Atheists

Rory J.Q. Barnes

An unbiased guide to the building blocks of spiritual belief.

Paperback: 978-1-78279-380-9 ebook: 978-1-78279-379-3

Readers of ebooks can buy or view any of these bestsellers by clicking on the live link in the title. Most titles are published in paperback and as an ebook. Paperbacks are available in traditional bookshops. Both print and ebook formats are available online.

Find more titles and sign up to our readers' newsletter at http://www.johnhuntpublishing.com/christianity
Follow us on Facebook at
https://www.facebook.com/ChristianAlternative